T0323668

Data, Statistics, and Useful Numbers for Environmental Sustainability

Data, Statistics, and Useful Numbers for Environmental Sustainability

Bringing the Numbers to Life

Benoit Cushman-Roisin
Bruna Tanaka Cremonini

Thayer School of Engineering,
Dartmouth College,
Hanover, NH, USA

ELSEVIER

Elsevier
Radarweg 29, PO Box 211, 1000 AE Amsterdam, Netherlands
The Boulevard, Langford Lane, Kidlington, Oxford OX5 1GB, United Kingdom
50 Hampshire Street, 5th Floor, Cambridge, MA 02139, United States

Notices
Knowledge and best practice in this field are constantly changing. As new research and experience broaden our understanding, changes in research methods, professional practices, or medical treatment may become necessary.

Practitioners and researchers must always rely on their own experience and knowledge in evaluating and using any information, methods, compounds, or experiments described herein. In using such information or methods they should be mindful of their own safety and the safety of others, including parties for whom they have a professional responsibility.

To the fullest extent of the law, neither the Publisher nor the authors, contributors, or editors, assume any liability for any injury and/or damage to persons or property as a matter of products liability, negligence or otherwise, or from any use or operation of any methods, products, instructions, or ideas contained in the material herein.

Library of Congress Cataloging-in-Publication Data
A catalog record for this book is available from the Library of Congress

British Library Cataloguing-in-Publication Data
A catalogue record for this book is available from the British Library

ISBN: 978-0-12-822958-3

For information on all Elsevier publications visit our website
at https://www.elsevier.com/books-and-journals

Publisher: Candice Janco
Acquisitions Editor: Marisa LaFleur
Editorial Project Manager: Aleksandra Packowska
Production Project Manager: Paul Prasad Chandramohan
Cover Designer: Christian J. Bilbow

Working together
to grow libraries in
developing countries

www.elsevier.com • www.bookaid.org

Typeset by TNQ Technologies

Contents

4. Pollutants and greenhouse gases

13. Making the numbers speak

Preface

In the context of environmental studies and sustainability, we frequently hear and read statements like these: "By installing low-flow showerheads, my school will save 1,250 gallons of water each year," or "this clean source of electricity will power 275,000 homes," or "it takes 75,000 trees to print a Sunday Edition of the Big-City Times." Don't you sometimes wish that you, too, could run the numbers and make similar statements about your environmental aspirations and accomplishments or about the impacts of others? All it takes are a few numbers that permit to translate paper usage into trees cut, electrical consumption into homes, recycling into energy, and energy savings into a number of miles not driven. But where are those numbers?

Many of these numbers exist but are scattered across scientific publications and Internet postings, some more reliable than others. It takes time to find the numbers, good numbers. At Dartmouth, the first author noted that his students spent a fair amount of time looking for these numbers in the pursuit of their assignments and projects, only to rediscover pretty much the same numbers course after course. So, to save time to his students, he began to compile a list of numbers, which he plainly titled "Useful Numbers" and posted it on one of his course websites. This list kept growing as years passed, and soon it grew to such a size that visitors to the website started asking questions: Where did you find this number? Do you have numbers about that? And it dawned on him that there was a definite need in the environmental community for an accessible and organized compilation of "useful numbers." The idea of gathering and organizing a relatively comprehensive set of numbers into a small handbook became a natural outgrowth.

And so, by popular demand, here is the handbook that you may have been looking for. It was prepared with diligence, from vetted information buried in refereed scientific publications as well as reliable websites. Effort has also been made to organize the disparate and often disconnected topics into some logical order. There is no right organization for this type of material, of course, and someone else would certainly have organized the topics differently. We the authors only hope that most users will find the present organization sufficiently practical.

This, obviously, is not a book that will ever be read from cover to cover, unless one wishes to use it for sleep therapy. It is a handbook where the user begins with the index pages in the back. Sources are cited and gathered at the end of each chapter, to keep them close to the numbers that they provided.

This little handbook should serve as a handy reference for researchers, teachers, and students to find quickly the number they seek in the preparation of their article, lecture, or assignment. Accessorily, it may also be useful for journalists and others who reach the broader public. Perhaps it can become the little book on the corner of your desk that, at the reach of the arm, puts meaning in your number.

Chapter 1

Materials

Materials extraction in the world in 2013 was estimated at 84.4 billion (10^9) metric tons (45.8% industrial and construction materials, 26.8% biomass, 17.2% fossil fuels, and 10.2% metals, excluding unused portion), or 11.8 metric tons per capita per year [1].

In the United States during 2013, materials consumption including fuels was estimated at 6.5 billion (10^9) metric tons per year, corresponding to 23.6 metric tons per capita per year, 51% higher than in Europe [2].

1.1 Metals

1.1.1 Aluminum

Aluminum production from bauxite via alumina is one of the most energy-intensive processes in the industry. According to a United Nations report based on 1979 data, the energy profile of aluminum is as tabulated below.

Energy consumption in the production of aluminum					
	Energy consumption		Energy type		
Stage	MJ/kg	Fraction	Electricity	Gas	Other
Mining	2.3	0.6%	6%	40%	54%
Alumina	44.5	17.1%	12%	80%	7.5%
Smelting	193.6	68.6%	85%	3%	12%
Fabrication	38.6	13.7%	38%	51%	11%
Total	279.0	100%			

Source: [3] Table 7 page 76.

Technologies have evolved ([4] page 125) toward greater energy efficiency. Compilation of multiple data sets for primary production (excluding the fabrication stage) yields an embodied energy in aluminum of 210 ± 10 MJ/kg ([4] page 471).

The carbon footprint for primary production is 12 kg of CO_{2eq}/kg whereas the water usage varies from 495 to 1,490 L/kg ([4] page 471).

Data, Statistics, and Useful Numbers for Environmental Sustainability.
https://doi.org/10.1016/B978-0-12-822958-3.00012-1

Recycling aluminum demands much less energy, only 26 MJ/kg for cast aluminum and 26.7 MJ/kg for wrought aluminum [5]. Its carbon footprint is 2.1 kg CO_2/kg (in average).

1.1.2 Copper and its alloys

According to a United Nations report using 1975 data, the energy profile of copper is as tabulated below. Although these numbers are dated, the breakdown is illustrative.

Energy consumption in the production of copper						
	Energy consumption		Energy type			
	MJ/kg	fraction	Electricity	Gas	Oil	Other
Mining	25.1	19.2%	42%	1%	35%	22%
Concentrating ore	49.1	37.7%	73%	6%	1%	20%
Smelting	44.3	34.0%	11%	48%	24%	17%
Refining	9.5	7.2%	43%	15%	39%	2%
Melting and casting	2.3	1.7%	—	—	—	—
Total	130.3	100%	—	—	—	—

Source: [3] Table 6 page 75.

Technologies have evolved, leading other sources to provide lower numbers.

Production impacts of copper and its alloys			
Metal	Embodied energy	Carbon footprint	Water usage
	(MJ/kg)	(kg CO_{2eq}/kg)	(L/kg)
Copper—pure Primary Recycled (40−45%)	57−63 12.9−14.3	3.7 1.07	293−324
Copper alloys Primary Recycled (40−45%)	56−62 12−15	3.7 0.83	268−297
Brass Primary Recycled (41−45%)	52−60 12−15	3.7 1.06	310−340
Bronze Primary Recycled (41−45%)	58.2−64.9 13.2−14.7	3.7 1.1	280−314

Source: [4]: page 477 and accompanying CES EduPack software.

1.1.3 Iron and Steel

The amount of energy consumed in steel production varies widely based on the process used and on the fraction of scrap metal with iron ore in the feed material. The Basic Oxygen Furnace consumes 23.2 MJ/kg while the Electric Arc Furnace consumes 9.3 MJ/kg [6]. The theoretical thermodynamic limit is 7.6 MJ/kg ([3] page 76).

In the United States, the energy consumed for the production of steel is about 19 MJ/kg from iron ore [3] and 10 MJ/kg from scrap metal ([3] page 76).

Almost 40% of the world's steel production is made from scrap [7]. Recycling 1 kg of steel saves 1.1 kg of iron ore, 0.63 kg of coal, 0.055 kg of limestone, 0.642 kWh of electricity, 0.287 L of oil, 10.9 thousand BTUs (=11.5 MJ) of energy, and 0.0023 m^3 of landfill [7].

Production impacts of iron and steel			
	Embodied energy	Carbon footprint	Water usage
Metal	(MJ/kg)	(kg CO$_{2eq}$/kg)	(L/kg)
Cast iron Primary Recycled (40−45%) Casting energy	16−20 10−11 10−11	1.5 0.52 0.79	13−39
Low-carbon steel Primary Recycled (40−44%)	25−28 6.6−8.0	1.8 0.44	23−69
Low-alloy steel Primary Recycled (40−44%)	31−34 7.7−9.5	2.0 0.52	37−111
Stainless steel Primary Recycled (35−40%)	81−88 11−13	5.0 0.73	112−336
Sources: [4] pages 463, 465, 467, and 469.			

1.1.4 Lead

Production of 1 kg of lead (Pb) from ore (galena, PbS) requires 27 MJ, 175−525 L of water and generates 2.0 kg of CO$_{2eq}$, while production of lead from recycled sources (mostly discarded automobile batteries) consumes 7.5 MJ/kg and generates 0.45 kg CO$_{2eq}$/kg. The single largest use of lead (70% of total production) is as electrodes in lead-acid batteries ([4] page 479).

1.1.5 Magnesium

Production of magnesium causes the following environmental impacts.

Production impacts of magnesium			
	Embodied energy	Carbon footprint	Water usage
Metal	(MJ/kg)	(kg CO_{2eq}/kg)	(L/kg)
Magnesium Primary Recycled (37−41%)	300−331 46−51	36.5 5.5	932−1,030
Magnesium alloys Primary Recycled (36−41%)	300−330 23−26	36 2.9	500−1,500
Source: [4]: page 473 and accompanying CES EduPack software.			

1.1.6 Nickel

Production of nickel causes the following environmental impacts.

Production impacts of nickel			
	Embodied Energy	Carbon Footprint	Water Usage
Metal	(MJ/kg)	(kg CO_{2eq}/kg)	(L/kg)
Nickel-chromium alloys Primary Recycled (29−32%)	173−190 30−36	11.5 2.0	564−620
Nickel-based superalloys Primary Recycled (22−26%)	221−244 33.8−37.5	11.6 2.14	134−484
Source: [4]: pages 483 and 485.			

Together with chromium and other elements, nickel is a component of stainless steel. For example, 18/8 stainless steel contains 18% chromium and 8% nickel while 18/10 is 18% chromium and 10% nickel.

1.1.7 Specialty and precious metals

Production impacts of specialty and precious metals			
	Embodied energy	Carbon footprint	Water usage
Metal	(MJ/kg)	(kg CO_{2eq}/kg)	(L/kg)
Gold Primary Recycled (42%)	240,000−265,000 650−719	26,500 43	126,000−378,000

Continued

Iridium Primary Recycled (0.7%)	43,000–47,600 2,000–2,210	2,900 165	186,000–206,000
Palladium Primary Recycled (3%)	149,000–165,000 5,140–5,680	8,500 426	186,000–206,000
Platinum Primary Recycled (3%)	257,000–284,000 7,760–8,570	14,750 642	186,000–206,000
Rhodium Primary Recycled (0.7%)	531,000–587,000 13,500–14,900	30,450 1,120	186,000–206,000
Silver Primary Recycled (66%)	1,400–1,550 140–170	100 9.3	1,150–3,460
Titanium alloys Primary Recycled (23%)	650–720 78–96	46.5 5.2	470–1,410

Source: [4]: pages 127, 474, 487, 488 and accompanying CES EduPack software.

1.1.8 Zinc

Production of zinc causes the following environmental impacts.

Production impacts of zinc			
	Embodied energy	Carbon footprint	Water usage
Metal	(MJ/kg)	(kg CO_{2eq}/kg)	(L/kg)
Zinc Primary Recycled (21–24%)	43.9–48.5 10.6–11.8	3.3 0.88	327–361
Zinc die-casting alloys Primary Recycled (21–25%)	57–63 10–12	4.1 0.67	160–521

Source: [4]: page 481.

1.2 Plastics and rubber

Except for biodegradable plastics, conventional plastics are made from so-called feedstock derived from crude oil refining and natural gas processing. The rule of thumb (borne by the numbers below) is that half the fossil fuel goes into the plastic itself while the remaining half is combusted to provide the energy during manufacture. Thus, it takes about 2 kg of fossil fuel to produce 1 kg of plastics. Since petroleum holds in average 43 MJ/kg, it takes approximately 86 MJ to produce 1 kg of plastics, and, with about 3 hydrogen atoms for every carbon atom in the fuel consumed in production (molar mass of 15 g per mole), the CO_2 emission (with a molar mass of 44 g per mole) is $44/15 = 2.9$ kg of CO_2 for every kg of plastics produced[1]. Actual amounts vary with the type of plastics, as the table below indicates.

Production impacts of plastics				
Polymers and elastomers	Acronym	Embodied energy (MJ/kg)	Greenhouse gas emission (kg CO_{2eq}/kg)	Water usage (L/kg)
Acrylonitrile butadiene styrene	ABS	90–99	3.6–4.0	250–277
Recycling		42–51	2.5–3.1	
Epoxy		127–140	6.8–7.5	107–322
Ethylene-vinyl-acetate	EVA	75–83	2.0–2.2	100–289
Recycling		42–52	2.5–3.1	
High-density polyethylene	HDPE	100–111	3.43–3.79	166–183
Recycling		26.2	0.90–0.99	
Phenolics		75–83	3.4–3.8	94–282
Polyamides (Nylons)	PA	116–129	7.6–8.3	250–280
Recycling		38–47	2.3–2.8	
Polycarbonate	PC	103–114	5.7–6.3	142–425
Recycling		38–47	2.3–2.8	
Polychloroprene	CR	61–68	1.6–1.8	126–378
Polyester		68–75	2.8–3.2	100–264
Polyethylene	PE	77–85	2.6–2.9	38–114
Recycling		45–55	2.7–3.0	
Polyethylene terephthalate	PET	81–89	3.7–4.1	14.7–44.2
Recycling		35–43	2.1–2.6	
Polyhydroxyalkanoate	PHA	81–90	4.1–4.6	100–300
Recycling		35–43	2.1–2.6	
Polylactide	PLA	49–54	3.4–3.8	100–300
Recycling		33–40	2.0–2.4	

Continued

1. *Note*: The amount of 6 kg of CO_2 emitted per kg of plastic mentioned by Time for Change [8] is inaccurate.

Polypropylene	PP	75−83	2.9−3.2	189−209
Recycling		45−55	2.0−2.2	
Polystyrene	PS	92−102	3.6−4.0	108−323
Recycling		43−52	2.6−3.1	
Polyurethane	PU/PUR	82.7−91.5	3.52−3.89	93.5−103
Recycling		28.1−31.1	1.2−1.32	
Polyvinyl chloride	PVC	56−62	2.4−2.6	77−85
Recycling		32−40	1.9−2.4	
Rubber—natural	NR	64−71	2.0−2.2	$(15-20) \times 10^3$
Rubber—butyl rubber	BR	112−124	6.3−6.9	63.8−191
Styrene		110−122	3.95−4.37	385−426

Source: [4]: Chapter 15 and accompanying CES EduPack software.

Recycling 1 kg of plastics saves 5.774 kWh of electricity, 2.604 L of oil, 98 thousand BTUs (=103 MJ) of energy, and 0.022 m^3 of landfill [7].

1.3 Paper and cardboard

In general, there can grow 16−20 mature trees on 1 acre (40−59 trees per hectare) [9], but the significantly smaller, softwood trees needed for the paper (so-called pulpwood) can be grown 12 feet apart from one another, for a density of 303 trees per acre (747 trees per hectare). Time from planting to harvest ranges from 7 to 10 years [10]. In regions where trees are grown for papermaking, the water needed for tree growth comes from natural precipitation, thus causing no environmental impact.

A cord of wood is 8ft × 4 ft × 4ft = 128 ft^3 and if air-dried and consisting of hardwoods weighs about 2 short tons (1,800 kg), about 15−20% of which is still water. One cord of wood makes 1,000−2,000 lbs of paper, depending on the process [11].

The production of 1 metric ton of paper requires 17 trees, in average, with the following spread: 24 trees for 1 ton of uncoated virgin (nonrecycled) printing and office paper but only 12 trees for 1 ton of 100% virgin (non-recycled) newsprint, 15.36 trees for 1 ton of higher-end magazine paper (for glossy magazines), and 7.68 trees for 1 ton of lower-end magazine paper (most catalogs) [12].

The production of 1 metric ton of paper consumes 51,500 MJ of energy, 25 m^3 of water, 680 gallons ($=2.57$ m^3) of oil and generates 1,150 kg of CO_{2eq} [4,12,13].

A "pallet" of copier paper[2] (20-lb. sheet weight) contains 40 cartons and weighs 1 metric ton. It contains 440 reams with each ream of paper containing 500 sheets and weighing 5 lbs ($=2.27$ kg). Therefore, the energy, water, and carbon footprints of a single page of paper are 234 kJ, 0.11 L water, and 5.23 g CO_2. It also follows that [12],

- 1 carton (10 reams) of 100% virgin copier paper uses 0.6 trees;
- 1 tree makes 16.67 reams of copy paper or 8,333 sheets;
- 1 ream (500 sheets) uses 6% of a tree;
- 1 ton of coated, higher-end virgin magazine paper (used for high-end magazines) uses 15.36 trees;
- 1 ton of coated, lower-end virgin magazine paper (used for newsmagazines and most catalogs) uses 7.68 trees.

In the United States, paper and cardboard recovery reached 66.8% in 2015. Of this, 33.4% went to produce corrugated cardboard, 11.8% noncorrugated cardboard (boxboard), 8.6% tissue, and 0.8% newsprint. Net exports accounted for 39.8%. Also, 36% of the fibers used to make new paper come from recycled sources [14].

1.4 Chemicals

In the United States, the chemical industry consumes an average of 6,935 BTUs per lb of product [15]. This energy intensity, however, depends widely on the nature of the chemical, as illustrated in the table on the following pages.

Because different production paths consume different amounts of energy, the energy used in the production of a chemical depends on its feedstock. Chemicals obtained from the cracking and distillation of petroleum or inorganic sources are called raw materials. Thus, the total energy consumed in the production of a chemical is the sum of the energy inputs for its production and that of all its predecessors (each with corresponding mass ratio deduced from the stoichiometric ratio), starting from the raw material. *Example*: The energy consumed in producing 1 lb of ethylene glycol from ethylene oxide (with mass ratio 0.710:1) from ethylene as raw material (with mass ratio 0.637:1) is:

$$E = 2{,}045 + 0.710 \times (1{,}711 + 0.637 \times 8{,}107) = 6{,}923 \text{ BTU/lb}.$$

To such number may be added the energy necessary for the intermediate production of the required hydrogen and chlorine.

---

2. Also called white office paper or printing and writing paper.

Energy consumption in the production of various chemicals			
Chemical	Energy consumption BTUs/lb unless otherwise noted	Made from	Mass ratio
Acetic acid (vinegar)	2,552 [16]		
Acetone	7,850		
Acrylonitrile	956	Propylene	0.793:1
Ammonia	12,150		
Ammonium	323	Ammonia	0.944:1
Ammonium nitrate	341	Nitric acid	0.787:1
Ammonium phosphate	323		
Ammonium sulfate	4,000	Ammonia	0.258:1
Benzene	1,255	Petroleum	
Bisphenol A (BPA)	6 MJ/kg [17]		
1,3-Butadiene	95	by-product of ethylene	
Carbon black	3,703 MJ/ton [18]		
Chlorine	4,800	Sodium chloride	1.648:1
Cumene (Isopropylbenzene)	696	Benzene	0.650:1
Cyclohexane	1,743	Benzene	0.928:1
Ethyl benzene	1,404	Benzene	0.736:1
Ethylene	8,107	Petroleum	
Ethylene dichloride	3,410	Ethylene	0.283:1
Ethylene glycol	2,045	Ethylene oxide	0.710:1
Ethylene oxide	1,711	Ethylene	0.637:1
Formaldehyde	150 kWh/ton [18]		
Hydrochloric acid	1.2 MJ/kg [18]		
Hydrogen	1.8 GJ/ton		
Isopropyl alcohol	4,693	Propylene	0.700:1
Methanol	38.4 GJ/ton		
Methyl tert-butyl ether (MTBE)	1,871 [19]		

Continued

Nitric acid	267	Ammonia	0.270:1
Oxygen	1.8 GJ/ton [19]		
Phenol and acetone together	7,850	Cumene	0.790:1
Phosphoric acid	4,300	Sulfuric acid	1.001:1
Polyethylene (PE)	1,178	Ethylene	1.000:1
Polypropylene	514	Propylene	1:000:1
Polystyrene	2,264	Styrene	1.000:1
Poly vinyl chloride (PVC)	1,246	Ethylene dichloride	
Propylene	1,351	Petroleum	
Propylene glycol	2,045	Propylene oxide	0.763:1
Propylene oxide	2,557	Propylene	0.725:1
Sodium carbonate	3,393		
Sodium chloride	Negligible	Sea salt	
Sodium hydroxide	3,765	Sodium chloride	1.461:1
Sodium silicate	5,344 MJ/ton [20]		
Sulfuric acid	1,047		
Styrene	16,891	Ethyl Benzene	1.019:1
Styrene butadiene (synthetic rubber)	2,271	1,3-Butadiene and Styrene	1.000:1
Terephthalic acid	1,779	Xylene	
Titanium dioxide	24.8 GJ/ton [20]		
Toluene	1,025	Petroleum	
Urea	843	Ammonia	0.567:1
Xylene	1,025		

Source: [15] based on 1997 data, unless otherwise noted.

1.5 Shaping of materials

Energy is not only spent in producing materials but also in shaping them into their ultimate useful shapes. The table below lists the most common processes.

Environmental impacts of shaping materials			
Material	Shaping process	Energy use (MJ/kg)	Carbon emission (kg CO_{2eq}/kg)
Metals[1]	Casting	8−12	0.4−0.6
	Rough, foil, rolling	3−5	0.15−0.25
	Extrusion, foil rolling	10−20	0.5−1.0
	Wire drawing	20−40	1.0−2.0
	Metal powder forming	20−30	1−1.5
	Vapor phase methods	40−60	2−3
Polymers	Extrusion	3.1−5.4	0.16−0.27
	Molding	11−27	0.55−1.4
Ceramics	Ceramic powder form	20−30	1−1.5
Glasses	Glass molding	2−4	0.1−0.2
Composites	Compression molding	11−16	1.6−2.3
	Spray/lay up	14−18	0.7−0.9
	Filament winding	2.7−4.0	0.14−0.2
	Autoclave molding	100−300	5−15

[1]*For variations across metals, see Ref. [4].*
Source: [4] page 133.

1.5.1 Primary shaping processes

Energy and carbon footprint of basic material processing techniques			
Process type	Variant	Energy use (MJ)	Carbon emission (kg of CO_{2eq})
Machining (per kg removed)	Heavy	0.8−2.5	0.06−0.17
	Finishing (light)	6−10	0.4−0.7
	Grinding	25−35	1.8−2.5
	Water jet, EDM, laser	500−5,000	35−350

Continued

Welding (per m welded)	Gas welding	1−2.8	0.055−0.15
	Electric welding	1.7−3.5	0.12−0.25
Fasteners (per fastener)	Fasteners, small	0.02−0.04	0.0015−0.003
	Fasteners, large	0.05−0.1	0.0037−0.0074
Adhesives (per m^2)	Cold	7−14	1.3−2.8
	Heat-curing	18−40	3.2−7.0
Painting (per m^2)	Solvent-based	50−60	0.63−0.95
	Baked coating	60−70	0.9−1.3
	Powder coatings	67−86	3.7−4.6
Plating (per m^2)	Electroplating	80−100	4.4−5.3

Source: [4] page 135.

1.5.2 Polymer shaping

Energy consumption and carbon footprint of shaping polymers				
	Molding		**Extrusion**	
Polymer	**Energy use (MJ/kg)**	**CO_2 footprint (kg/kg)**	**Energy use (MJ/kg)**	**CO_2 footprint (kg/kg)**
Acrylonitrile butadiene styrene (ABS)	18−20	1.4−1.5	5.8−6.4	0.44−0.48
Polyamides (Nylons, PA)	21−23	1.55−1.7	5.9−6.5	0.44−0.49
Polypropylene (PP)	20.4−22.6	1.5−1.7	5.9−6.5	0.44−0.49
Polyethylene (PE)	22.7−25.1	1.7−1.9	6.0−6.6	0.45−0.49
Polycarbonate (PC)	17−6.19.5	1.3−1.5	5.8−6.4	0.43−0.48
Polyethylene terephthalate (PET)	18.7−20.6	1.4−1.55	5.8−6.4	0.44−0.48
Polyvinylchloride (PVC)	13.9−15.4	1.05−1.16	5.6−6.3	0.42−0.47
Polystyrene (PS)	16.5−18.3	1.24−1.37	5.7−6.4	0.43−0.48
Polyhydroxyalka-noates (PHA, PHB)	16.6−18.4	1.25−1.38	5.8−6.4	0.43−0.48
Polylactide (PLA)	15.4−17	1.15−1.27	5.7−6.3	0.43−0.47

Continued

Epoxy		21−23	1.7−1.85		
Polyester		26−28	2.1−2.3		
Phenolics		26−29	2.1−2.3		
Rubber	Natural	15−17	1.2−1.4		
	Butyl	14−16	1.2−1.4		
Ethylene-Vinyl-Acetate (EVA)		13.8−15.2	1.1−1.2	5.4−6.0	0.43−0.48
Polychloroprene (Neoprene, CR)		17.2−18.5	1.37−1.5		
Source: [4] pages 492−525.					

1.6 Miscellaneous materials

Energy and carbon footprints of miscellaneous materials					
		Primary production			
Material		**Embodied energy (MJ/kg)**	**Carbon footprint (kg CO_{2eq}/kg)**	**Water usage (L/kg)**	
Automotive	Anti-freeze	76			
	Engine oil	60.2			
	Other fluids	52			
Carbon fiber		450−500	33−36	360−1,367	
Ceramics	Alumina	49.5−54.7	2.67−2.95	29.4−88.1	
	Glass	10−11	0.7−0.8	14−20.5	
	Glass-recycling	7.4−9.0	0.44−0.54		
	Glass - laminated	27.7−30.6	1.67−1.84	28.7−31.8	
	Pyrex	27−30	1.6−1.8	26−37.5	
	Pyrex - recycling	20−23	1.2−1.4	26−37.5	
Construction	Brick	2.2−3.5	0.20−0.23	2.8−8.4	
	Cement		0.927		
	Concrete[1]	1.0−1.3	0.09−0.12	1.7−5.1	
	Sand	1.0			
	Stone	0.4−0.6	0.03−0.04	1.7−5.1	

Continued

Cotton		44–48	2.4–2.7	7,400–8,200
Fiberglass (glass-fiber reinforced plastic-GFRP)		107–118	7.47–8.26	105–309
Foams	Rigid polymer	96–107	3.7–4.1	299–865
	Flexible polymer	104–114	4.3–4.7	181–544
Straw bale		0.1–0.3	−1.1–0.9	0
Wood	Bamboo flooring[2]	1.23	−0.0168	0
	Hardwood	9.8–10.9	0.8–0.94	500–750
	Hardwood during construction	0.46–0.55	0.022–0.027	0
	Plywood	13–16	0.78–0.87	500–1,000
	Plywood during construction	0.455–0.55	0.022–0.027	0
	Softwood	8.8–9.7	0.36–0.40	500–750
	Softwood during construction	0.46–0.55	0.022–0.027	0
Wool		51–56	3.2–3.5	$(1.6–1.8) \times 10^5$

[1] Including 10% of cement in the concrete mix.
[2] Bamboo flooring has a negative carbon footprint because of the emissions during its production amount to less than the amount sequestered by the plant [21].
Sources: [4] Chapters 5, 15 & 20, [21] for bamboo, Fig. 2.

A significant fraction of the CO_2 emitted during the production of cement is reabsorbed into the concrete over the course of its life cycle, in a process called carbonation. A study estimates that 33–57% of the CO_2 emitted during cement production will be absorbed through the carbonation of concrete surfaces over a 100-year life span [22].

Sources

[1] Sustainable Europe Research institute (SERI) in cooperation with Vienna University of Economics and Business. www.materialflows.net/fileadmin/docs/materialflows.net/WU_MFA_Technical_report_2015.1_final.pdf. See also Organization for Economic Cooperation and Development (OECD). www.oecd.org/greengrowth/MATERIAL%20RESOURCES,%20PRODUCTIVITY%20AND%20THE%20ENVIRONMENT_key%20findings.pdf

[2] University of Michigan – Center for Sustainable Systems – U.S, Material Use Factsheet, Pub. No. CSS05-18, 2019. For older data, see U.S. Geological Survey, Fact Sheet 2009-3008, css.umich.edu/sites/default/files/US%20Material%20Use_CSS05-18_e2019.pdf, pubs.usgs.gov/fs/2009/3008/

[3]   United Nations Center on Transnational Corporations — Climate Change and Transnational Corporations — Analysis and Trends. U. N. Centre on Transnational Corporations, Environ. Ser. 2 (1992). ST/CTC/112, ISBN 92-1-104385-9, Chapter 7 "Production of Energy Intensive Metals", 110 pages, ieer.org/wp/wp-content/uploads/1992/01/ClimateChange-TransnationalCorp_1992-FULL.pdf.

[4]   M.F. Ashby, Materials and the Environment — Eco-Informed Material Choice, second ed., Butterworth-Heinemann, 2013, 616 pages.

[5]   M. Schubert, K. Saur, H. Florin, P. Eyerer, H. Beddies, Life Cycle Analysis — Getting the Total Picture on Vehicle Engineering Alternatives, Automotive Engineering, March 1996, pp. 49—52.

[6]   U.S. Department of Energy, Energy Information Administration (eia) — 2011: Steel Industry Analysis Brief — Energy Consumption. www.eia.doe.gov/emeu/mecs/iab98/steel/intensity.html.

[7]   Bureau of International Recycling, The Industry. www.bir.org/industry/.

[8]   Time for Change, Plastic bags and plastic bottles — CO_2 emissions during their lifetime. timeforchange.org/plastic-bags-and-plastic-bottles-CO2-emissions.

[9]   Wood M., Let Your Trees Grow For Profit (undated article). www.woodmagazine.com/materials-guide/lumber/let-your-trees-grow-for-profit.

[10] How it's Made, Forest for Paper (About Sappi Company in South Africa). howitsmade.co.za/growing-forests-for-paper-pulp/.

[11] Sierra Club, How much paper does one tree produce? www.sierraclub.org/sierra/2014-4-july-august/green-life/how-much-paper-does-one-tree-produce.

[12] Conservatree.org, Trees into Paper — How Much Paper Can Be Made from a Tree? conservatree.org/learn/EnviroIssues/TreeStats.shtml.

[13] C. Thompson, Recycled Papers — the Essential Guide, MIT Press, Cambridge, MA, 1992, 200 pages. Quoted by, conservatree.org/learn/EnviroIssues/TreeStats.shtml.

[14] Paper Recycles. paperrecycles.org/statistics/paper-paperboard-recovery and. paperrecycles.org/statistics/where-recovered-paper-goes.

[15] U.S. Department of Energy, Office of Industrial Technologies — Energy and Environmental Profile of the U.S. Chemical Industry, May 2000. www1.eere.energy.gov/manufacturing/resources/chemicals/pdfs/profile_full.pdf.

[16] M. Neelis, E. Worrell, E. Masanet, Energy Efficiency Improvement and Cost Saving Opportunities for the Petrochemical Industry, Lawrence Berkeley National Laboratory, June 2008. www.energystar.gov/ia/business/industry/Petrochemical_Industry.pdf.

[17] R. Agrawal, A. Suman, Production of Bisphenol A, Dept. Chemical Engineering, Jaypee University of Engineering and Technology, Guna, India, 2012. www.scribd.com/doc/94377374/Production-of-Bisphenol-A#scribd.

[18] H.-J. Althaus, R. Hischier, M. Osses, A. Primas, S. Hellweg, N. Jungbluth, M. Chudacoff, Life Cycle Inventories of Chemicals, Swiss Center for Life Cycle Inventories, Zürich, Ecoinvent Report No. 8, 2007. December 2007, 957 pages, db.ecoinvent.org/reports/08_Chemicals.pdf.

[19] E. Worrell, D. Phylipsen, D. Einstein, N. Martin, Energy Use and Energy Intensity of the U.S. Chemical Industry, University of California Berkeley, 2000. LBNL-44314, April 2000, 34 pages, escholarship.org/content/qt2925w8g6/qt2925w8g6.pdf.

[20] European Commission — Integrated Pollution Prevention and Control — Large Volume Inorganic Chemicals — Solids and Others Industry, August 2007, 711 pages, eippcb.jrc.ec.europa.eu/sites/default/files/2019-11/lvic-s_bref_0907.pdf.

[21] L. Gu, Y. Zhou, T. Mei, G. Zhou, L. Xu, Carbon footprint of bamboo scrimber flooring − implications for carbon sequestration of bamboo forests and its products, Forests, MDPI 10 (2019) 51−64, doi.org/10.3390/f10010051.

[22] C. Pade, et al., The CO_2 Uptake of concrete in the Perspective of Life Cycle Inventory. International Symposium on Sustainability in the Cement and Concrete Industry, Lillehammer, Norway, September 2007. Quoted on page 8 of [20], 2007.

Chapter 2

Water

2.1 Hydrological cycle

Every year, $456 \times 10^{12}\,\mathrm{m^3}$ of water evaporates from the ocean, which corresponds to the removal of the upper 1.27 m of water from the ocean surface, while $62 \times 10^{12}\,\mathrm{m^3}$ of water is removed from land areas by a combination of evaporation and transpiration. Annual precipitation is $410 \times 10^{12}\,\mathrm{m^3}$ over the oceans and $108 \times 10^{12}\,\mathrm{m^3}$ over land, causing a net transfer of $46 \times 10^{12}\,\mathrm{m^3}$ of water from the oceans via the atmosphere and return of the same amount of water from the land to the ocean by surface and subsurface runoff ([1] page 355).

The total amount of water on earth is estimated to be $1.386 \times 10^9\,\mathrm{km^3}$, of which only 2.5% ($3.5 \times 10^7\,\mathrm{km^3}$) is in the form of freshwater. The amounts in various forms and their estimated residence times are tabulated as follows.

Distribution of water on the planet			
Location	Amount (km³)	Fraction	Residence time
Oceans	1,338,000,000	96.5%	3,200 years
Polar ice caps, glaciers, permanent snow	24,064,000	1.74%	20–20,000 years
Groundwater	23,400,000	1.69%	100–10,000 years
Ground ice and permafrost	300,000	2.2×10^{-4}	–
Freshwater lakes	91,000	6.6×10^{-5}	50–100 years
Saline lakes	85,400	6.2×10^{-5}	–
Soil moisture	16,500	1.2×10^{-5}	1–2 months
Atmospheric water vapor	12,900	9.3×10^{-6}	9 days
Rivers and swamps	2,120	1.5×10^{-6}	2–6 months
Living biomass	1,120	8.1×10^{-7}	2 weeks in human body

Sources: [1–3].

Data, Statistics, and Useful Numbers for Environmental Sustainability.
https://doi.org/10.1016/B978-0-12-822958-3.00009-1

2.2 Energy for water

Energy is required for the distribution of municipal drinking water, specifically for its extraction in nature and conveyance to the treatment plant, the various treatment processes, and the pumping through the often crusty pipe network of the municipality.

If the water source is from an underground aquifer, the pumping energy requirement increases with depth: 537 kWh per million gallons (0.142 kWh/m^3) from a depth of 120 ft (36.6 m), 896 kWh per million gallons from a depth of 200 ft (122 m), which corresponds to about 70% pumping efficiency ([4] page 11). The other energy requirements are tabulated below, with the higher numbers being typical for California where water procurement and treatment are more complicated.

Energy required to handle water		
Step	**Energy required (kWh/10^6 gallons)**	**Energy required (kWh/1,000 m^3)**
Source and conveyance	6,260 0–14,000	1,654 0–3,700
Treatment	184 100–16,000	49 26.5–4,200
Distribution	1,013 700–1,200	268 185–320
Total	7,460 800–31,200	1,970 211–8,220

Sources: [4] page 32, [5] page 2.

Energy is required to extract freshwater from seawater. For a typical salinity of 3.45% salt and at a temperature of 25°C, the minimum energy required is 0.86 kWh per m^3 of freshwater produced [6]. In practice, the amount of energy required far exceeds this minimum and varies depending on the desalination process.

Energy required for desalination of water by various processes				
	Multistage flash distillation (MSF)	MED-TVC combination	Multieffect distillation (MED)	Mechanical vapor compression (MVC)	Reverse osmosis (RO)
Capacity (m^3/day)	50,000 –70,000	10,000 –35,000	5,000 –15,000	100 –2,500	24,000
Electricity consumption (kWh/m^3)	4–6	1.5–2.5	1.5–2.5	7–12	3–5.5

Thermal energy consumption (kWh/m³)	9.5—19.5	9.5—25.5	5—8.5	None	None
Total energy consumption (kWh/m³)	13.5—25.5	11—28	6.5—11	7—12	3—5.5

Source: [6], crediting Wangnick Consulting (2010).

2.3 Water consumption

While the amounts of water withdrawals for various activities vary with the state of the economy and because of growing population and affluence, the following numbers for the United States during 2005 ought to be representative, at least as far as their proportions are concerned.

Water volumes handled in the United States in 2005				
		Amount used		
Activity		**(10^6 L/day)**	**(10^6 gal/day)**	**Fraction of freshwater**
Power plants	Freshwater	541,000	143,000	41%
	Seawater	219,000	58,000	—
	All	760,000	201,000	—
Irrigation		484,000	128,000	37%
Public supply		167,100	44,200	12.7%
Industrial uses		68,800	18,200	5.2%
Aquaculture		33,200	8,780	2.5%
Mining		15,200	4,020	1.2%
Livestock		8,100	2,140	0.6%
Other withdrawals		2,500	660	<1%
Totals	Freshwater	1,320,000	349,000	100%
	Seawater	230,600	61,000	—
	All	1,550,000	410,000	—

Source: [7] page 1.

The freshwater footprint per person in the United States (total consumption above divided by population of 295.5 million in 2005) is 1,200 gallons/day (=4,500 L/day).

Domestic consumption of water in the United States, excluding outdoor use (which varies greatly with location and residence type) is estimated to break down as tabulated below.

Breakdown of domestic water consumption in the United States			
	Amount used per person		
Activity	(L/day)	(gal/day)	Fraction
Toilets	70	18.5	26.7%
Laundry	57	15.0	21.7%
Showers	44	11.6	16.8%
Faucets	41	10.9	15.7%
Leaks	36	9.5	13.7%
Bathing	4.5	1.2	1.7%
Dishwashing	3.7	1.0	1.4%
Other domestic uses	6.0	1.6	2.3%
Total	262	69.3	100%

Source: [1] page 365.

For water usage in the processing of materials, the reader is referred to the individual materials listed in Chapter 1.

Humans also consume water indirectly, some of which is through food production as the table below indicates.

Water consumption in the production of selected foods		
	Amount of water used in production per unit weight	
Food	L/kg	gallons/lb
Beef	45,510	5,214
Pork	13,600	1,630
Chicken	6,800	815
Apples	409	49
Carrots	275	33
Wheat	210	25
Potatoes	200	24
Tomatoes	190	23
Lettuce	190	23

Source: [8].

In large buildings where cooling towers are used to provide central air conditioning, the rate of water consumption is 3 gallons of water per minute for each ton of refrigeration[1] provided [9], which amounts to 0.015 gallons of water per BTU of heat removed. In metric units, this amounts to 54 L of water per MJ of heat removed. For the average commercial building relying on cooling towers to provide 500 tons of refrigeration, the water consumption is 1,500 gallons of water per minute totaling 900,000 gallons for a 10-hour day, which is equivalent to 13,000 occupants each consuming 69.3 gallons per day (table above). Put another way, the refrigeration of the average commercial building by means of cooling towers consumes in a single 10-hour day as much water as one person consumes in 13,000 days = 36 years.

Below is the breakdown of water usage in restrooms of large buildings. Not surprisingly, most of the water is used to flush toilets.

Water usage in restrooms			
Toilets	Urinals	Showers	Faucets
72%	17%	7%	4%
Source: [10] Fig. 4–3.			

2.4 Wastewater

In the United States, domestic wastewater generation is about 120 gallons per person per day (=450 L/day), with a spread of 50–250 gallons per person per day ([11] page 314).

Typical wastewater flowrates from various locations and activities are as tabulated below.

Water consumption in various types of locations				
		Flow (L/day)		
Source		Typical	Range	Unit
Airport		11	8–15	per passenger
Apartment	High rise	190	130–280	per person
	Low rise	245	280–300	per person
	Resort	230	190–265	per person

Continued

---

1. A ton of refrigeration is a unit of power equal to 12,000 BTUs/hr = 3,516 W.

Automobile service station		38	26–49	per vehicle served
		45	34–57	per employee
Bar		11	4–19	per customer
		49	38–61	per employee
Cabin, resort		150	30–190	per person
Cafeteria		8	4–11	per customer
		38	30–45	per employee
Campground (with facilities)		115	75–150	per person
Cocktail lounge		75	45–95	per seat
Coffee shop		23	15–30	per customer
		38	30–45	per employee
Cottage (summer use)		150	95–190	per person
Country club		380	230–490	per member present
		49	38–57	per employee
Day camp—no meals		49	38–57	per person
Department store		1,890	1,510 –2,270	per toilet room
		38	30–45	per employee
Dining hall		26	15–38	per meal served
Dormitory, bunkhouse		150	75–190	per person
Hospital—medical		627	475–910	per bed
		38	19–57	per employee
Hospital—mental		380	285–530	per bed
		38	19–57	per employee
Hotel		182	150–214	per guest
		38	26–49	per employee
House	Typical home	265	170–340	per person
	Better home	300	230–380	per person
	Luxury home	360	285–570	per person
	Older home	170	115–220	per person
Industrial building (toilets only)		49	26–61	per employee

		2,080	1,700 −2,460	per laundry machine
Laundry facility (self-service)		190	170−210	per wash
Motel	with kitchen	380	340−680	per unit
	without kitchen	360	285−570	per unit
Office building		49	26−61	per employee
Prison		435	285−570	per inmate
		38	19−57	per employee
Rest home		320	190−455	per resident
Restaurant		11	8−15	per meal served
School	without cafeteria or gym	42	19−64	per student
	with cafeteria	57	38−75	per student
	with cafeteria and gym	95	57−115	per student
	boarding	285	190−380	per student
Shopping center (mall)		38	26−49	per employee
		8	4−8	per parking space
Store (retail)		11	4−15	per customer
		38	30−45	per employee
Swimming pool		38	19−45	per customer
		38	30−45	per employee
Theater		11	8−15	per seat
Trailer park		150	115−190	per person
Visitor center		19	15−30	per visitor

Add contributions for places with multiple groups of people, such as those with both customers and employees.
Source: [12].

Sources

[1] J.R. Mihelcic, J.B. Zimmerman, Environmental Engineering − Fundamentals, Sustainability, Design, John Wiley & Sons, 2010, 695 pages.

[2] U.S. Geological Survey − Where is Earth's Water?

[3] PhysicalGeography.net. www.physicalgeography.net/fundamentals/8b.html.

[4] R. Cohen, B. Nelson, G. Wolff, Energy Down the Drain: The Hidden Costs of California's Water Supply, National Resources Defense Council, Pacific Institute, Oakland, California, August 2004. www.nrdc.org/sites/default/files/edrain.pdf.

[5] U.S. Environmental Protection Agency, Energy Efficiency in Water and Wastewater Facilities, 2013. epa.gov/statelocalclimate/documents/pdf/wastewater-guide.pdf.

[6] DESWARE. Encyclopedia of Desalination and Water Resources. www.desware.net/desa4.aspx.

[7] J.F. Kenny, N.L. Barber, S.S. Hutson, K.S. Linsey, J.K. Lovelace, M.A. Maupin, Estimated Use of Water in the United States in 2005, U.S. Geological Survey, Circular 1344, 2009. pubs.usgs.gov/circ/1344/pdf/c1344.pdf.

[8] J. Robbins, The Food Revolution, tenth ed., Conari Press, 2010, 480 pages. For more accessible quotes from this book, see Vegetarianism and the Environment at, michaelbluejay.com/veg/environment.html.

[9] Conservation Mechanical Systems, Inc. Water Use in Cooling Towers. www.conservationmechsys.com/wp-content/siteimages/Water_use_in_Cooling_Towers.pdf.

[10] P.H. Gleick, et al. (Eds.), Waste Not, Want Not: The Potential for Urban Water Conservation in California, Pacific Institute, November 2003, 165 pages. pacinst.org/wp-content/uploads/2003/11/waste_not_want_not_full_report.pdf.

[11] M.J. Hammer, M.J. Hammer Jr., Water and Wastewater Technology, fourth ed., Prentice Hall, 2001, 636 pages.

[12] G. Tchobanoglous, F.L. Burton, H.D. Stensel, Wastewater Engineering − Treatment and Reuse, Metcalf and Eddy, McGraw-Hill, 2002, 1848 pages. Table adaptation by Timothy G. Ellis, Dept. Civil, Construction & Environmental Engineering, Iowa State U., as posted at, www.eolss.net/EolssSampleChapters/C06/E6-13-04-05/E6-13-04-05-TXT-05.aspx#4._Wastewater_Quantities_.

Chapter 3

Energy

3.1 Units

Energy is probably the physical quantity with the most diverse set of units at our disposal, allowing us to quantify a wide range of amounts from the tiny energy of a single electron to the amount consumed annually in the world. Units also tend to differ depending on the context, not being the same for electricity, petroleum, food, thermal engines, and buildings. Below are the main units in alphabetical order [1]:

1 BOE (barrel-of-oil-equivalent) = 5.80×10^6 BTUs = 6.1179×10^9 J = 1,699 kWh

1 BTU (British Thermal Unit) = 1,055.056 J = 2.92875×10^{-4} kWh = 251.996 cal

1 cal (calorie) = 0.003968 BTUs = 4.186 J (see food calorie below)

1 EJ (exa-joule) = 10^{18} J = 9.478×10^{14} BTUs

1 erg = the work of 1 dyne over 1 cm = 10^{-7} J = 6.2415×10^{11} eV

1 eV (electron-volt) = 1.60218×10^{-19} J

1 GJ (giga-joule) = 10^9 J = 947,820 BTUs = 277.78 kWh

1 Gtoe (giga-tonne of oil equivalent) = 1,000 Mtoe = 41.868 EJ = 39.68 quad

1 J (joule) = the work of 1 Newton moving 1 meter = 0.00094782 BTUs = 0.2389 cal

1 kJ (kilo-joule) = 1,000 J = 0.94782 BTUs = 238.84 cal

1 kcal (kilo-calorie) = 1 food calorie = 1,000 cal = 3.9683 BTUs = 4,186.8 J

1 kWh (kilo-watt-hour) = 3,412.1 BTUs = 3.600 MJ = 859.9 kcal

1 lb-steam (pound of steam) = 970 BTUs

1 MJ (mega-joule) = 10^6 J = 947.82 BTUs = 0.27778 kWh

1 MMBTU = 10^6 BTUs = 1.055056×10^9 J = 293.07 kWh

1 Mtoe (million tonne of oil equivalent) = 4.1868×10^{16} J = 1.163×10^7 MWh

1 MWh (mega-watt-hour) = 1,000 kWh = 3,600 MJ

1 Q (quad) = 10^{15} BTUs = 1.055056 EJ

1 Therm = 100,000 BTUs

1 Wh (watt-hour) = 3.412 BTUs = 3.600 kJ = 859.9 cal

Data, Statistics, and Useful Numbers for Environmental Sustainability.
https://doi.org/10.1016/B978-0-12-822958-3.00001-7

Power, which is energy per time, has some units of its own [2]:

1 hp (horsepower) = 2,542.47 BTUs/hour = 745.70 W
1 W (watt) = 1 J/s = 3.4121 BTUs/hour = 1.341 hp
1 kW (kilo-watt) = 1,000 W
1 MW (mega-watt) = 10^6 W
1 GW (giga-watt) = 10^9 W
1 Ton (ton of refrigeration) = 12,000 BTUs/hour = 3,516 W

3.2 Solar energy

The temperature of the sun is 5,778K = 5,505°C.

The solar radiation shining at the top of the atmosphere is 1,367 W/m^2 ($\pm$3% as the earth orbits the sun) [3] and spans a spectrum from about 200 nm (nanometer) to 4,000 nm, with a peak around 550 nm, consisting of 10% ultraviolet (UV), 40% visible light, and 50% infrared (IR) [4]. Visible light has wavelengths from 380 (red) to 780 nm (violet) [4].

After adjustment for the angle of incidence and lack of radiation at night, the solar flux averaged over the earth's surface is 340 W/m^2, of which 48% or 163 W/m^2 reaches the earth's surface [5]. A "peak-sun" is defined as 1,000 W/m^2, which is about what a mid-latitude location receives at noon when the sky is clear.

The amount of solar energy reaching the earth every hour (628 EJ) is greater than the amount of energy used by the human population over an entire year (389 EJ in 2013 [6]).

3.3 Energy generation

From Albert Einstein, we know that mass can theoretically be converted into energy according to $E = mc^2$. This formula provides a theoretical maximum for energy extraction per mass: $E/m = c^2 = 89.9 \times 10^9$ MJ/kg. Needless to say, fuels used by humans fall far short of this theoretical limit, as the numbers in the following indicate.

Of the total primary energy supply in the world in 2014, 81% is from carbon-based fossil sources divided into oil (31%), coal (29%), and natural gas (21%). Next sources are biofuels & waste (10.1%), nuclear (4.8%), hydro (2.4%), and other renewables (1.3%) [6].

3.3.1 Solid fuels

The amount of thermal energy released by the combustion of the most common fuels is tabulated as follows.

Energy released by combustion			
Material		**MJ/kg**	**BTUs/lb**
Cardboard		16—19	6,900—8,200
Coal	Anthracite	30.1	12,190
	Bituminous coal	25.0—33.4	10,750—14,340
	Charcoal	28	12,000
	Lignite ("brown coal")	16.1	6,910
	Peat	21	9,000
Nuclear	^{235}U in theory[1]	83,140,000	3.57×10^{10}
	^{235}U in burner reactor	500,000	2.15×10^{8}
Paper		16—19	6,900—8,200
Plastics	Polyethylene	45	19,400
	Polypropylene	45	19,400
	Polystyrene	40	17,200
	PVC (bottles, etc.)	15—25	6,500—10,700
Solid waste	Municipal garbage	20	8,600
	Animal dung	10	4,300
	Dried bagasse	16	6,900
	Food scraps	15—20	6,500—8,600
	Rice husks	16	6,900
	Shelled corn		392,000 BTUs/bushel[2]
Switch grass	(oven dried)	18.0	7,750
Wood	Green (50% MC)[3]	10.0	4,300
	Semidried (30% MC)[3]	14.0	6,020
	Air-dried (20% MC)[3]	16.0	6,880
	Oven dried (0% MC)[3]	20.0	8,600
	Seasoned		20×10^{6} BTUs/cord[4]
	Wood pellets	15.8	6,800
	Premium wood pellets	19.1	8,200

[1] The fission of 1 atom of Uranium-235 generates 3.24×10^{-11} J, which amounts to 83.14 TJ/kg.
[2] 1 bushel = 64 US pints = 35.2 L.
[3] MC = Moisture Content, on wet basis.
[4] A cord of wood is 8 ft × 4 ft × 4 ft, including spaces between logs. A typical woodstove is only 65—75% efficient; a pellet stove is 83% efficient; the rest of the heat goes out the chimney with the smoke.
Sources: [7—11].

In the United States in 2013, 80 waste-to-energy plants incinerated 30 million tons of municipal solid waste and generated 14 billion kWh of electricity, about the same amount used by 1.3 million households [12].

3.3.2 Liquid fuels

One barrel of oil holds 42 US gallons (=0.159 m^3), weighs 136 kg, and has an energy content of 5.80 × 10^6 BTUs (=6.12 GJ) [13]. Equivalently, 1 metric ton of crude oil holds 42.65 × 10^6 BTUs (=45.00 GJ) in the form of chemical energy.

Below are the energy contents of the most common liquid fuels.

Energy released by combustion			
Fuel	in MJ/kg	in MJ/L	in BTUs/gallon
Biodiesel	40.5	35.6	128,000
Crude oil	45	38.4	138,000
Diesel	50	38.7	139,000
Ethanol	30	23.6	84,530
Fuel oil #2	40	36.7	138,800
Fuel oil #6 ("Bunker oil")	45	41.8	150,000
Gasoline ("Petrol")	48	34.6	124,000
Kerosene	44	37.6	135,000
Liquefied natural gas (LNG)	55	–	23,700 BTUs/lb
Liquefied propane (LPG)	46	25.5	91,330
Methanol	23	18.2	65,200

Sources: [9,10,13−15].

3.3.3 Gaseous fuels

Since gases are compressible, their energy content is best quoted on a per-mass basis.

Energy released by combustion		
Fuel	in MJ/kg	in BTUs/lb
Biogas	45	19,300
Butane	50.3	21,640
Hydrogen	142.1	61,084

Methane	55.4	23,811
Natural gas	45.4−52.3	19,500−22,500
Propane	50.2	21,564

Sources: [9,16].

On a volume basis at ambient temperature, natural gas holds 1,025 BTUs/ft^3 (=38.2 MJ/m^3) [10]. This makes 10^6 ft^3 (=28,320 m^3) of natural gas equivalent to 177 barrels of crude oil. At standard compression for delivery trucks, propane holds 91,300 BTUs/gallon (=25.4 MJ/L) [10].

3.3.4 Compost

Compost can be used to heat or pre-heat water for domestic consumption. A compost heap generates heat at the rate of 1,500 BTUs/h per short ton of material (=1,740 J/hr/kg) for the first 4 weeks and 500 BTUs/hr per short ton (=580 J/hr/kg) for the following 4 weeks ([17] page 118). The volume to mass ratio is 2 cubic yards per short ton (=1.7 m^3 per metric ton). Typical temperatures inside a compost heap are 120−130°F (50−55°C) toward the bottom of the heap and 150−180°F (66−82°C) toward the top (because hot air rises), regardless of the outdoor atmospheric temperature ([17] page 117).

The outputs from 1 metric ton (2,205 lbs, including 50% by weight) of compost are 3.375 MBTUs, 290 kg of CO_2 and 47 L of water vapor over the course of a 21-day thermophilic stabilization period. These are equivalent to 1,530 BTUs, 0.29 lb CO_2, and 0.0056 gallons of water vapor, per pound of compost [18].

3.3.5 Animal manure

A composting system designed for small farms or homesteads claims an output of 50,000 BTUs/hr generated from 30 to 50 short tons of animal manure made available every 8 weeks, the equivalent of 12−20 cows or horses ([17] page 145).

In composting farmyard manure, bacteria generate 4.03 kWh per kg of oxygen consumed (=6,240 BTUS/lb O_2 = 14.5 MJ/kg O_2) [19].

3.3.6 Other biological matter

Thermophilic bacteria generate 4 Wh per gram of oxygen used (=6,190 BTUs per pound of O_2 = 14.4 MJ per kg of O_2) [20].

3.3.7 Photovoltaic cells

There are many different types of photovoltaic (PV) cells available on the market, and each type has its own efficiency, as shown in Figure 3-1. The 2015 efficiency of the typical crystalline silicon PV cell used in roof-top applications is 12−15% installed [21].

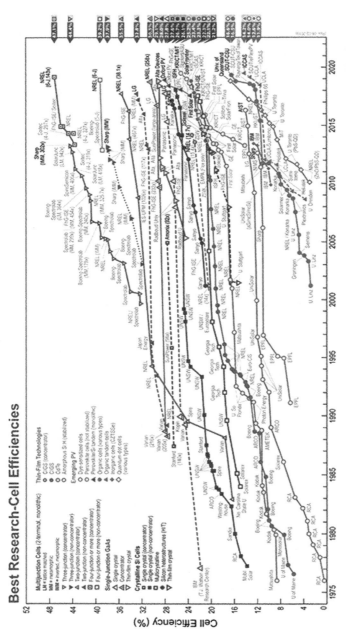

FIGURE 3-1 Confirmed conversion efficiencies for a variety of photovoltaic technologies, from 1976 to 2019, at a reference temperature of 25°C. *Source:* [21].

The energy output of a photovoltaic cell depends on various factors. Considering a final 12% conversion efficiency, the average annual energy output of a photovoltaic cell is 216 kWh/m^2 per year in the continental United States [22], 79 kWh/m^2 in Alaska, and 280 kWh/m^2 in the Sahara ([23] page 343).

3.3.8 Wind turbines

The following table provides the wind power density (Watts per square meter of area swept by the turbine blades) as a function of wind speed [24]. In wind energy, it has become traditional to ascribe "wind power class" numbers to various wind speed intervals based on height above the ground (=level of turbine nacelle). Physics dictates that the power in the wind increases like the cube of the wind speed.

Wind power according to wind class						
	10 m		30 m		50 m	
Wind power class	Power density (W/m^2)	Speed (m/s)	Power density (W/m^2)	Speed (m/s)	Power density (W/m^2)	Speed (m/s)
1	0–100	0–4.4	0–160	0–5.1	0–200	0–5.6
2	100–150	4.4–5.1	160–240	5.1–5.8	200–300	5.6–6.4
3	150–200	5.1–5.6	240–320	5.8–6.5	300–400	6.4–7.0
4	200–250	5.6–6.0	320–400	6.5–7.0	400–500	7.0–7.5
5	250–300	6.0–6.4	400–480	7.0–7.4	500–600	7.5–8.0
6	300–400	6.4–7.0	480–640	7.4–8.2	600–800	8.0–8.8
7	400–1,000	7.0–9.4	640–1,600	8.2–11.0	800–2,000	8.8–11.9

Source: [24].

Wind turbines are rated according to their capacity, which is the maximum power they can generate. At winds below a threshold called the cut-in wind speed, turbines do not turn. Above cut-in wind speed, their power generation ramps up with wind speed and eventually levels off (see example in Figure 3-2). The wind speed at which this leveling off occurs is called the rated wind speed. At high winds, above a threshold called the cut-out wind speed, turbines are forced to stop for fear of damage. Thus, they only generate their rated power in winds between rated and cut-out wind speeds. The variability of the local wind ultimately determines the power output as a percentage of the rated power of the turbine. This is called the capacity factor, and its value typically lies between 30 and 45%, with an average of about 40% [25].

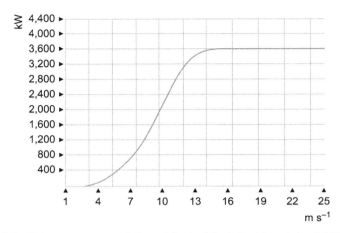

FIGURE 3-2   Power output versus wind speed for the GE wind turbine rated at 3.6 MW, with a rotor diameter of 111 m. The cut-in wind speed is 3.5 m/s, the rated wind speed is 14 m/s, and the cut-out wind speed is 27 m/s. *Source:* [26].

Onshore and offshore wind potential for the 10 countries with the largest carbon emissions is tabulated in the following.

Wind Energy Potential (TWh)			
Country	Onshore	Offshore	Total
Canada	78,000	21,000	99,000
China	39,000	4,600	44,000
Germany	3,200	900	4,100
India	2,900	1,100	4,000
Iran	5,600	–	5,600
Japan	570	2,700	3,200
Russia	120,000	23,000	140,000
Saudi Arabia	3,000	–	3,000
South Korea	130	990	1,100
United States	74,000	14,000	88,000

Source: [26].

A question in wind farm design is how much land is required for wind turbines. The rule of thumb is that turbines must be at least 7 rotor diameters away from one another in the direction of the wind and 5 rotor diameters away

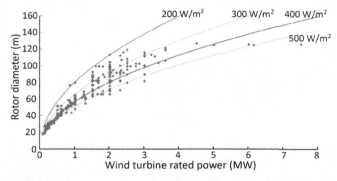

FIGURE 3-3   Relation between rated capacity and rotor diameter in wind turbines. *Source:* [28].

in the transverse direction [27]. This implies that the minimum horizontal footprint of one turbine is 35 rotor diameters squared. Rotor diameter D (in meters) is related to rated power output P (in MW) as shown in Figure 3-3, with the approximate rule $D = 55\sqrt{P}$, which implies that the maximum power capacity per unit area of land, wind strength permitting, is

$$\frac{P}{35D^2} = 9\frac{\text{MW}}{\text{km}^2}.$$

3.4 Energy conversion

3.4.1 Electricity generation

A traditional thermal power plant is 33.2% efficient and produces 1 kWh of electricity from 10,272 BTUs of thermal energy [29], which amounts to about 1 GWyr (giga-watt-year) for every quad of fuel.

In the United States, the generation of 1 kWh of electricity necessitates 95 L of water in average, with a low of 0.038 L/kWh for electricity from natural gas to a high of about 420 L/kWh for electricity from biodiesel [30]. See Section 4.3.4 for the carbon footprint of electricity generation.

3.4.2 Internal combustion engine

The internal combustion engine (ICE), propelling most cars, trucks, and boats, is about 30% efficient [31]. For an automobile, after subtraction of parasitic losses (such as alternator, water pump, headlights, drivetrain losses, idling, etc.), the energy delivered to the wheels to move the vehicle is only 18–25% of the chemical energy held in the fuel, as Figure 3-4 illustrates.

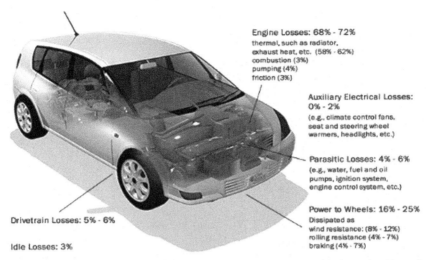

Engine Losses: 68% - 72%
thermal, such as radiator,
exhaust heat, etc. (58% - 62%)
combustion (3%)
pumping (4%)
friction (3%)

**Auxiliary Electrical Losses:
0% - 2%**
(e.g., climate control fans,
seat and steering wheel
warmers, headlights, etc.)

Parasitic Losses: 4% - 6%
(e.g., water, fuel and oil
pumps, ignition system,
engine control system, etc.)

Power to Wheels: 16% - 25%
Dissipated as
wind resistance: (8% - 12%)
rolling resistance (4% - 7%)
braking (4% - 7%)

Drivetrain Losses: 5% - 6%

Idle Losses: 3%

FIGURE 3-4 Where the energy goes in a car after the fuel has combusted in the engine. For each attribution, the two different percentages refer to city and highway driving, respectively. *Source:* [31].

A 2004 study published in *Science* [32] calculates a tank-to-wheel efficiency of 12.6% for a 1,500 kg ICE car (lower than the range quoted earlier) and 27.2% for the hybrid-ICE car (above the range quoted earlier).

When the internal combustion engine is used as an electric generator (for off-grid applications or during emergencies, for example), it consumes 10,403 BTUs of petroleum fuel per kWh of electricity generated [33].

3.4.3 Electric motor/Alternator

In contrast to the internal combustion engine, the electric motor is far more efficient, at 87%, with the most powerful ones exceeding 90% [33]. When used in reverse as an alternator, the efficiency is about the same.

The company Tesla claims on its specs page that the 270/310 kW (=362/ 416 hp) AC induction electric motor in its 85 kWh Model-S automobiles is 93% efficient [34], and the more recent Model 3 uses permanent magnet reluctance motors, which achieve an efficiency of 97% [34]. After drivetrain and parasitic losses, this translates into 0.25 kWh per mile of driving at 50 mph (=80 km/h) [35]. However, one needs to keep in mind that the generation of electricity upstream of the car is a very inefficient process (see Section 3.3) and that the efficiency of an electric motor depends on its speed and torque.

3.4.4 Fuel cells

The proton-exchange-membrane (PEM) fuel cell, which produces electricity from hydrogen at ambient temperatures, is about 40−50% efficient [32,36]. A PEM fuel-cell car has an estimated tank-to-wheel efficiency of 26.6% [32].

The following table compares the efficiencies of various fuel-cell technologies.

Fuel cell efficiencies		
Fuel cell technology	Efficiency	With cogeneration of heat
Alkaline	60–70%	
Direct methanol	<40%	
Molten carbonate	65%	>85%
Phosphoric acid	37–42%	>85%
Proton exchange membrane (PEM)	40–50%	
Solid oxide	60–70%	up to 85%
Sources: [32,36,37].		

3.4.5 Biomass to ethanol

Ethanol can be produced from a variety of organic materials (biomass). The following table recapitulates the theoretical yields for the most common feedstocks. For conversion from volume to mass, use ethanol density of 0.789 kg/L.

Theoretical ethanol yield from various biomass sources		
Feedstock	Gallons per dry ton	L/kg
Bagasse	111.5	0.465
Corn grain	124.4	0.519
Corn stover	113.0	0.472
Cotton gin trash	56.8	0.237
Forest thinnings	81.5	0.340
Hardwood sawdust	100.8	0.421
Mixed paper	116.2	0.485
Rice straw	109.9	0.459
Switch grass	96.7	0.404
Wood	100.2	0.418
Source: [38].		

Ethanol can also be produced from algae in a body of water exposed to sunlight and atmospheric carbon dioxide. An experimental program reported producing 9,000 gallons of ethanol per acre per year (=84 m^3 per hectare per year) [39].

Fermentation of sugar in sweet beets yields 1,000 gallons of ethanol per acre based on a harvest of 35 short tons per acre (=78 metric tons per hectare) [40].

3.4.6 Oil crops to biodiesel

Biodiesel is a renewable fuel that can be produced from naturally oily plants, particularly oil palm. Oil palm trees produce 20 metric tons of fresh fruit per hectare per year, of which the oils form 10% of the total dry biomass while the remaining 90% may serve as a source of fiber or cellulosic material for additional biofuel production [41]. The following table compares the yield from various crops, including oil palm, which has the highest yield.

Biodiesel yield from various oil crops		
	Yield	
Plant	(L/hectare)	(gallons/acre)
Coconut	2,160	231
Oil palm	4,800	520
Peanut	820	88
Rapeseed	936	100
Soybean	526	56
Sunflower	760	81
Source: [41].		

3.4.7 Efficiency factors

The following table recapitulates the approximate efficiency factors for various types of energy conversion.

Efficiency of various types of energy conversion				
From	To	By means of	Efficiency	Source
	Electricity	Gasification + gas turbine + generator	20%	[42]
Biomass	Gas	Integrated gasification	45%	[42]
	Heat	Woodstove (with smoke up chimney)	65–83%	[43]

Coal	Electricity	Power plant via steam	33−37%	[44,45]
	Heat	Home coal furnace (with vented smoke)	55%	[46]
Diesel	Electricity	Stationary diesel generator	40%	[47]
	Mechanical	Automotive diesel combustion engine	40−48%	[45]
Electricity	Heat	Electric resistance	100%	[46]
	Hydrogen	Electrolysis	81%	[44]
	Light[1]	Incandescent bulb	5%	[46]
		Fluorescent lamp	20−25%	[46]
		Light-emitting diode (LED)	50%	[48]
	Mechanical	Electric motor	80−93%	[33]
Fossil fuels	Heat	Combustion	100%	−
		Combustion (with vented smoke)	65%	[46]
Gasoline	Electricity	Generator	18−20%	[48]
	Mechanical	Internal combustion engine	30%	[31]
Hydro	Electricity	Water turbine + generator	93%	[47]
Hydrogen	Electricity	PEM fuel cell	40−50%	[32,44]
		Phosphoric acid fuel cell	65%	[37]
Mechanical	Electricity	Generator/alternator	90−95%	[46]
Natural gas	Mechanical	Gas turbine	50−60%	[47]
	Electricity	Gas turbine + generator	50%	[47]
	Heat	Home gas furnace	85%	[46]
Nuclear	Electricity	Uranium fission + steam cycle	33%	[47]
Sunlight	Biomass	Photosynthesis	3−6%	[49]
	Electricity	Photovoltaic cell	3−22%	[21]
Steam	Heat	Boiler	85%	[46]
	Mechanical	Steam turbine	45	[34]
Wind	Electricity	Wind turbine + generator	25−35%	[44]

[1]Based on 200 lumens = 1 W of emission in the visible spectrum.

3.5 Energy storage

3.5.1 Batteries

Three quantities best describe the capability of a battery: Its energy density (measuring how much energy can be stored per mass of battery, sometimes also called specific energy), its power density (measuring how quickly the energy can be retrieved from the battery), and its round-trip efficiency (ratio of electricity recovered to electricity stored, also called Coulombic efficiency). The following table compares these numbers for a variety of commercially available batteries.

Technical characteristics of different battery types			
Battery	Energy density (Wh/kg)	Power density (W/kg)	Round-trip efficiency
Lead-acid	35	180	>80%
Lithium-ion	118–225	200–430	>95%
Lithium-ion Polymer	130–225	260–450	>91%
Nickel-cadmium	50–80	200	75%
Nickel-metal Hydride	70–95	200–300	70%
Redox flow	10–50	>50	85%
Sodium-nickel chloride	90–120	155	80%
Sodium-sulfur	150–240	150–230	80%
Vanadium redox flow	50	110	–
Zinc-bromine	65–75	90–100	88–95%
Zinc-air	294–442	100	–

Sources: [50]-Table 1, [51]-Table 37.2, [52]-Table 1, [53].

For comparison, the energy density of gasoline is 12,890 Wh/kg [13], then reduced to 3,900 Wh/kg at the shaft because of the low 30% efficiency of the internal combustion engine [31], and its power density easily exceeds 1,000 W/kg (=horsepower rating of the engine divided by its mass). The round-trip efficiency of gasoline is 0% because the engine is incapable of converting mechanical energy back into fuel energy.

The 85-kWh lithium-ion battery set on board the 2012 Tesla Model S is reported having the following characteristics [54]: 140 Wh/kg energy density, 516 W/kg power density, and 75% round-trip efficiency.

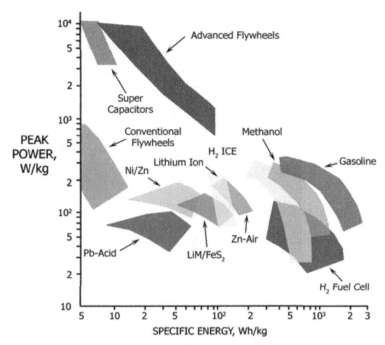

FIGURE 3-5   The Ragone Plot displaying various energy storage technologies according to their holding capacity (specific energy, in Wh/kg) and their retrieval rate (peak power, in W/kg). *Source:* [55]-Fig. 38.

Figure 3-5, called the *Ragone Plot*, displays graphically the comparison between various types of energy storage: commercially available batteries, supercapacitors, flywheels, fuel cells, and fossil fuels, based on energy density and peak power. Note the logarithmic scale of each axis.

3.5.2 Hydrogen

Being a gas that can be relatively easily generated and consumed, hydrogen may be used as a form of energy storage. Since its chemical energy content (see Section 3.3.3) is 142.1 MJ/kg (=39.5 kWh/kg), any hydrogen system will necessarily have an energy density lower than 39.5 kWh/kg.

For most applications, hydrogen may not be stored at ambient pressure but must be greatly compressed, liquefied, or stored in a metal hydride. Compression and liquefaction entail a significant energy penalty whereas storage in a metal hydride adds mass to the system. Thus, either way, the energy density of a complete system falls far below the theoretical maximum. The following table shows the energy density of systems as of 2006, per mass and per volume.

Energy density of various hydrogen storage technologies		
	Energy density	
Technology	(kWh/kg)	(kWh/L)
Compressed to 5,000 psi (34.5 MPa)	1.9	0.5
Compressed to 10,000 psi (69 MPa)	1.6	0.8
Liquefied	1.7	1.2
Stored in metal hydride	0.8	0.6
Source: [56].		

For hydrogen produced by electrolysis and converted back to electricity, the round-trip efficiency is the efficiency of electrolysis times that of a proton-exchange membrane fuel cell, that is, $0.81 \times 0.45 = 36\%$. By contrast, the round-trip efficiency of storing hydrogen in a metal hydride ($Mg + H_{2gas} \leftrightarrow MgH_2$) is around 74% [57] and can reach 90% for storage in metal-organic frameworks (MOFs) [58].

3.5.3 Pumped hydro

Pumped hydro entails the pumping of water from a lower to a higher reservoir during energy storage and the release of this water back to the lower reservoir through a water turbine during energy recovery. The technology is mature and permits the storage of large quantities of energy but is applicable only where land is suitable and when energy release occurs over hours or days. Some systems have been in existence since the 1920s.

According to a 2013 Sandia National Laboratories report ([59] page 33), systems can be sized up to 4,000 MW and operate with round-trip efficiencies of 76–85%. A reservoir 25 m deep and 1 km in diameter situated 200 m higher than another reservoir (or natural lake) can hold enough water to generate 10,000 MWh.

3.5.4 Compressed air

Compressing air to store energy entails a certain amount of adiabatic temperature increase depending on the rapidity of compression. A subsequent cooling that returns the temperature to ambient value during the storage period causes a heat loss, which may not be recovered during expansion. There is also some mechanical inefficiency in the compressor itself.

A system storing air at 4,500 psi (=31,030 kPa) above atmospheric pressure can hold 22–49 Wh/L ([53] page 14), which is much less than batteries, with a theoretical round-trip efficiency ranging from 8% for isentropic

(=rapid) compression and expansion to 72% for a three-stage polytropic compression and three-stage polytropic expansion ([60] page 27), and up to 82% with great care ([60] page 19).

3.6 Energy transport

3.6.1 Electrical transmission

Electricity is easily transported by electrically conductive wires. Depending on the line voltage, two stages are distinguished: transmission (at high voltage over long distances) and distribution (at low voltage across communities). A percentage of loss is quoted for the combination of transmission and distribution. Values depend on the quality of the infrastructure and the geographical distances in the country.

Efficiency of electricity of transmission according to location		
Region	**Country**	**Losses**
North America	Canada	8.6%
	Mexico	14.3%
	United States	6.0%
Latin America	Argentina	16.0%
	Brazil	16.4%
	Cuba	15.4%
	Haiti	54.2%
Latin America and Caribbean combined		14.8%
Europe	Belgium	4.9%
	Denmark	5.5%
	Central Europe and Baltics	7.5%
	Finland	3.7%
	Germany	3.9%
	Italy	7.4%
	Netherlands	4.4%
	Norway	8.0%
	Spain	9.0%
	United Kingdom	7.5%

Continued

	Algeria	18.4%
Africa	Kenya	18.0%
	Nigeria	15.3%
	Senegal	16.0%
	South Africa	8.5%
	Sub-Saharan Africa	11.8%
Middle East	Arab World	11.9%
	Israel	4.0%
	Saudi Arabia	7.0%
	Middle East and North Africa	12.0%
Central Asia	Russia	10.1%
South Asia	India	18.5%
	Overall	18.1%
East Asia	China	5.8%
	Japan	4.6%
	East Asia and Pacific Islands	6.1%
Oceania	Australia	5.9%
	New Zealand	6.7%
World average		8.2%

Source: [61] from 2013 data.

3.6.2 Pipelines and oil tankers

Conveying petroleum or natural gas by means of a pipeline over long distances requires pumping and thus entails an energy penalty. Pressure drop is caused by friction along the inner pipe walls as well as in bends, expansions, and contractions because of secondary motions.

Along natural gas pipelines, compressors that operate on a nonstop basis are needed every 50–100 miles (80–160 km) to maintain adequate pressure. The average pumping station, utilizing multiple compressors for a total of 9,984 hp (=7.45 MW) in average per station, is capable of moving 461 million cubic feet of natural gas per day (=13.1 × 10^6 m^3/day) [62]. This amounts to 21.6 hp per million cubic feet of gas moved per day (=0.57 W per (m^3 per day)). Assuming that the average distance between consecutive pumping stations is 75 miles (=121 km), the power necessary to keep the gas moving per unit distance is 0.29 hp per million cubic feet conveyed per day over 1 mile, or 4.7 kW per million m^3 conveyed per day over 1 km. In terms of energy expenditure per distance per mass of gas, the number is 1.7 MJ/km per metric ton of gas conveyed in the pipeline [63].

The 764 km long (=475 miles) Iraq Crude Oil Pipeline connecting the South Rumaila oil fields to the Red Sea has six intermediate pumping stations each equipped with 3 turbopumps driven by 22 MW turbines, for a total maximum pumping power of 396 MW [64]. Assuming that this maximum power is consumed only when the pipeline is utilized at full capacity of 1.6 million barrels per day (=2.94 m^3/s), the power consumption is estimated at 176 kW per m^3/s conveyed over 1 km. Further taking that crude oil has an energy content of 38.4 MJ/L (Section 3.3.2), the relative energy loss in pipeline transportation is 0.46% per 1,000 km (meaning that it takes 0.46 J to carry 100 J worth of crude oil over 1,000 km). In terms of energy expenditure per distance per mass of oil, the number is 0.2 MJ/km per metric ton of oil conveyed in the pipeline [63].

Crude oil is also transported across the oceans by oil tankers. Using 0.16 MJ of energy needed to transport 1 metric ton over 1 km by means of an ocean ship ([65] page 142) and the fact that crude oil contains 45 MJ per kg (Section 3.3.2), it is found that the energy efficiency of ocean shipping is 0.36% per 1,000 km (meaning that it takes 0.36 J to carry 100 J worth of crude oil over 1,000 km).

3.7 Energy consumption

3.7.1 Transportation

Energy for transportation is quantified as the energy needed to convey a certain number of passengers or tonnage of goods for a certain distance by a certain mode of transport. The numbers can be found in Chapter 5 together with the corresponding greenhouse gas emissions.

3.7.2 Buildings

In the United States, it was estimated in 2014 that the electricity consumption in a residential home was 10,932 kWh per year in average[1], ranging from a low of 6,077 kWh/yr in Hawaii (where the climate is very mild) to a high of 15,497 kWh/yr in Louisiana (where much electricity is used in summer for air conditioning) [66]. Since energy (kWh) divided by time (year) is power, this amounts to: 1.25 kW/home for the nation, 1.77 kW/home in Louisiana, and 0.69 kW/home in Hawaii.

For large-scale energy projects such as concentrated solar systems or wind farms, an accurate translation of power generated into a number of homes being powered should take into account the fact that approximately 6% of the

---

1. Beware of the numbers used by some environmental organizations, in newspapers and magazines, and in advertisements, for they are often skewed.

electrical power is lost in transmission and distribution [67]. This means that, because of transmission and distribution losses, 6% fewer homes can actually be powered.

3.7.3 Food

The relation between food and energy is twofold: (1) Food is a form of energy through its caloric content, and (2) significant amounts of energy are required in the production of food, starting with the manufacturing of fertilizers and tilling of the land to processing, packaging, storing, and transportation at various stages of food production. The following data are taken from Ref. [68] and references therein.

Accounting for a 26% waste of edible food, the average American consumed 2,590 calories per day (=10.84 MJ/day) in 2010. The ratio of energy used for the production of food to the amount of energy in the food is 7.36 to 1 (10.3 quads for food production compared to 1.4 quads in the food, in the United States in 1999), which amounts to a meager conversion efficiency of $1/7.36 = 14\%$. Consumption of energy for food-related activities accounts for nearly 15% of the US energy budget.

The breakdown of energy use in the US food system is as follows.

Energy use in the US food system			
Activity		**Percentage of energy consumption**	
Agricultural production	Manufacturing of fertilizers and pesticides	8.6%	21.4%
	Farm activities	12.8%	
Transportation		13.6%	
Processing		16.4%	
Packaging materials		6.6%	
Food retail		3.7%	
Commercial food services		6.6%	
Household storage and preparation	Refrigeration	13.0%	31.7%
	Preparation	18.7%	
Total		100%	

Source: [68].

3.7.4 Human activities

The average human resting metabolic rate is 1,300 kilocalories per day (63 W), 20% (10.8 kcal per hour, 12 W) of which is consumed by brain activity [69]. The following table lists the energy expenditure of a 160 lbs (73 kg) individual for typical activities.

Energy expenditure by the human body			
Activity		**in kcal/hr**	**in W**
Brain activity		10.8	12.6
Sleeping		49	57
Sitting		89	104
Bowling		177	206
Bicycling	leisurely (<10 mph)	236	340
	racing (>20 mph)	944	1,098
Walking normally		211	245
Walking briskly (at 3.5 mph, 5.6 km/h)		314	365
Running		472	550
Skiing, downhill		314	365
Aerobics	low-impact	295	343
	general	384	447
	high impact	413	480
	in water	402	468
Tennis playing		413	480
Hiking		438	509
Martial arts		590	686
Swimming laps		423	492
Rope jumping		600	700
Running	at 5 mph (8 km/h)	606	705
	at 8 mph (13 km/h)	860	1,000
Sources: [69—72].			

Sources

[1] The Engineering ToolBox — Energy. www.engineeringtoolbox.com/unit-converter-d_185. html#Energy.

[2] The Engineering ToolBox — Power. www.engineeringtoolbox.com/unit-converter-d_185. html#Power.

[3] University of Oregon, Solar Radiation Monitoring Laboratory. solardat.uoregon.edu/ SolarData.html.

[4] U.S. National Renewable Energy Laboratory (NREL), Reference Solar Spectral Irradiance. rredc.nrel.gov/solar/spectra/am1.5/.

[5] NASA, Earth Observatory. earthobservatory.nasa.gov/Features/EnergyBalance/page4.php.

[6] International Energy Agency, World Balance (2013). www.iea.org/Sankey/#? c=World&s=Balance. International Energy Agency — Key World Energy Statistics, 2015. www.iea.org/publications/freepublications/publication/KeyWorld_Statistics_2015. pdf. International Energy Agency — Renewable energy continuing to increase market share, July 2016. www.iea.org/newsroomandevents/news/2016/july/renewable-energy-continuing-to-increase-market-share.html.

[7] U.S. Lawrence Berkeley National Laboratory, Energy and Power. muller.lbl.gov/teaching/ Physics10/PffP_textbook/PffP-01-energy-2008.pdf.

[8] The Engineering Toolbox, Standard Grade Coal and Heating Values. www.engineeringtoolbox. com/coal-heating-values-d_1675.html.

[9] United Nations Joint Logistic Center, Cooking Fuel Options Help Guide, Undated. stoves. bioenergylists.org/files/cooking_fuel.pdf.

[10] U.S. Department of Agriculture, U.S. Forest Service — Forest Products Laboratory, July 2004, 3 pages, www.yumpu.com/en/document/read/4242161/fuel-value-calculator-forest-products-laboratory-usda-forest-.

[11] World Bank, Municipal Solid Waste Incineration, undated. www.worldbank.org/urban/ solid_wm/erm/CWG%20folder/Waste%20Incineration.pdf.

[12] U.S. Energy Information Administration (eia), Biomass Explained — Waste-to-Energy (Municipal Solid Waste). www.eia.gov/energyexplained/?page=biomass_waste_to_energy.

[13] American Physical Society, Energy Units (and sources therein). www.aps.org/policy/ reports/popa-reports/energy/units.

[14] The Engineering Tool Box, Energy Content in Common Energy Sources. www. engineeringtoolbox.com/energy-content-d_868.html The Engineering Toolbox — Fuels Densities and Specific Volumes. www.engineeringtoolbox.com/fuels-densities-specific-volumes-d_166.html.

[15] U.S. Department of Energy, Alternative Fuels Data Center — Fuel Properties Comparison. www.afdc.energy.gov/fuels/fuel_comparison_chart.pdf.

[16] The Engineering Toolbox, Fuel Gases and Heating Values. www.engineeringtoolbox.com/ heating-values-fuel-gases-d_823.html.

[17] G. Brown, The Compost-Powered Water Heater, The Countryman Press, Woodstock, Vermont, 2014, 162 pages.

[18] R. Sardinsky, Greenhouse CO_2 dynamics and composting in a solar heated bioshelter, in: J. Hayes, D. Jaehne (Eds.), Solar Greenhouses: Living and Growing, Proceedings of the Second Conference on Energy Conserving and Solar Heated Greenhouses, American Solar Energy Society, Newark, New Jersey, 1979, pp. 22—40.

[19] I.F. Svoboda, M.R. Evans, Heat from aerated liquid animal wastes, in: E. Stentiford (Ed.), Proceedings of the First International Conference on the Composting of Solid Wastes and Slurries, 28—30 September 1983, University of Leeds, Dept. of Mechanical Engineering, Leeds, UK, 1983.

[20] C.L. Cooney, D.I.C. Wang, R.I. Mateles, Measurement of heat evolution and correlation with oxygen consumption during microbial growth, Biotech. Bioeng. 9 (1968) 269−281.

[21] U.S. National Research Energy Laboratory, National Center for Photovoltaics − Best Research-Cell Efficiencies, 2015. www.nrel.gov/pv/assets/pdfs/best-research-cell-efficiencies.20190802.pdf.

[22] U.S. National Renewable Energy Laboratory, PV FAQS - what Is the Energy Payback for PV?, 2004. www.nrel.gov/docs/fy04osti/35489.pdf.

[23] R. Toossi, Energy and the Environment − Choices & Challenges in a Changing World, third ed., Global Digital Press, 2014, 569 pages.

[24] J.F. Manwell, J.G. McGowan, A.L. Rogers, Wind Energy Explained: Theory, Design and Application, second ed., Wiley, 2010, 704 pages.

[25] Energy Numbers, What Does the Capacity Factor of Wind Mean? energynumbers.info/capacity-factor-of-wind.

[26] X. Lu, M.B. McElroy, Global potential for wind-generated electricity. Chapter 4, in: T.M. Letcher (Ed.), 2017: Wind Energy Engineering: A Handbook for Onshore and Offshore Wind Turbines, Academic Press, 2017, 622 pages.

[27] P.R. Mitchell, Wind Turbine Separation Distances Mater, North American Platform Against Windpower, 2014. June 2014, www.na-paw.org/Mitchell/Mitchell-Wind-Turbine-Separation-Distances.pdf.

[28] J. Serrano-González, R. Lacal-Arántegui, Technological evolution of onshore wind turbines—a market-based analysis, Wind Energy 19 (2016) 2171−2187. doi.org/10.1002/we.1974.

[29] U.S. Energy Information Administration, Average Tested Heat Rates by Prime Mover and Energy Source (2007−2014). www.gov/electricity/annual/html/epa_08_02.html. with relative proportions of the fuels from. www.eia.gov/energyexplained/index.cfm?page=electricity_in_the_united_states.

[30] W.D. Jones, How Much Water Does It Take to Make Electricity? IEEE Spectrum, 2008, 1 April 2008 (corrected 12 September 2011), spectrum.ieee.org/energy/environment/how-water-does-it-take-to-make-electricity.

[31] U.S. Dept. of Energy, Fuel Economy Information − Where the Energy Goes: Gasoline Vehicles. www.fueleconomy.gov/feg/atv.shtml.

[32] N. Demirdöven, J. Deutch, Hybrid cars now, fuel cell cars later, Science 305 (2004) 974-976. doi.org/10.1126/science.1093965.

[33] The Engineering Toolbox, Electrical Motor Efficiency. www.engineeringtoolbox.com/electrical-motor-efficiency-d_655.

[34] M. Humphries, Tesla Model S/X Upgrading to More Efficient Electric Motor. PC Magazine, (April 8, 2019). www.pcmag.com/news/367661/report-tesla-model-s-x-upgrading-to-more-efficient-electric. F. Lambert, Tesla Is Upgrading Model S/X with New, More Efficient Electric Motors. Electrek, (April 5, 2019). electrek.co/2019/04/05/tesla-model-s-new-electric-motors.

[35] Tesla, Model S Efficiency and Range, by Elon Musk and J. B. Straubel, May 9, 2012. www.tesla.com/blog/model-s-efficiency-and-range?redirect=no.

[36] HydrogenTrade.com, Hydrogen Fuel Cells. www.hydrogentrade.com/fuel-cells/.

[37] U.S. Department of Energy, Office of Energy Efficiency & Renewable Energy − Types of Fuel Cells. energy.gov/eere/fuelcells/types-fuel-cells.

[38] U.S. Department of Energy, Energy Efficiency and Renewable Energy − Alternative Fuel Center − Ethanol Feedstocks. www.afdc.energy.gov/fuels/ethanol_feedstocks.html.

[39] U.S. Department of Energy, Office of Energy Efficiency & Renewable Energy — Making Algal Biofuel Production More Efficient, Less Expensive. www.energy.gov/eere/articles/making-algal-biofuel-production-more-efficient-less-expensive.

[40] Renewable Energy World, Ethanol from Energy Beets: A Viable Option? by B. Dorminey, April 2014. www.renewableenergyworld.com/articles/print/volume-17/issue-2/bioenergy/ethanol-from-energy- beets-a-viable-option.html.

[41] United Nations Environment Programme, Environment for Development — Oil palm Plantations: Threats and Opportunities for Tropical Ecosystems, December 2011. na.unep.net/geas/getuneppagewitharticleidscript.php?article_id=73.

[42] U.S. Department of Energy, Industrial Technologies Division — Production of Electricity from Biomass Crops, by Ralph P. Overend (Undated). www.mtholyoke.edu/courses/tmillett/course/geog_304B/7290.pdf.

[43] U.S. Department of Energy, Home Heating Systems — Wood Pellet Heating. energy.gov/energysaver/wood-and-pellet-heating.

[44] J. Randolph, G.M. Masters, Energy for Sustainability — Technology, Planning, Policy, Island Press, 2008, 791 pages.

[45] R.A. Giannelli, E. Nam, Medium and Heavy Duty Diesel Vehicle Modeling Using a Fuel Consumption Methodology, U.S. EPA National Vehicle and Fuel Emissions Laboratory, Ann Arbor, Michigan, 2004, 27 pages, www3.epa.gov/otaq/models/ngm/may04/crc0304c.pdf.

[46] Pennsylvania State University — Prof, Ljubisa Radovic — Lecture Notes - Efficiency of Energy Conversion. www.ems.psu.edu/~radovic/Chapter4.pdf.

[47] Independent Energy, LLC, Energy Efficiency. www.independentenergyllc.com/Efficiency.html.

[48] Design Recycle, Inc, Comparison Chart LED Lights vs. Incandescent Light Bulbs vs. CFLs. www.designrecycleinc.com/led%20comp%20chart.html.

[49] United Nations, Food and Agriculture Organization — Renewable Biological Systems for Alternative Sustainable Energy Production — Chapter 1: Biological Energy Production, by K. Miyamoto (FAO Agricultural Services Bulletin — 128). www.fao.org/docrep/w7241e/w7241e05.htm#1.2.1 photosynthetic efficiency.

[50] N. Zimmerman, Vanadium Redox Flow Battery, Society & Engineering Degree Project, School of Business, Mälardalen University (Sweden), 2014. July 2014, 59 pages, www.diva-portal.se/smash/get/diva2:772090/FULLTEXT01.pdf. See Table 1 on Page 4 and references cited in footnotes.

[51] P.C. Butler, P.A. Eidler, P.G. Grimes, S.E. Klassen, R.C. Miles, Zinc/Bromine Batteries, Sandia National Laboratories, 2000, 16 pages, www.sandia.gov/ess/publications/SAND2000-0893.pdf.

[52] Duracell — Zinc-Air Battery, Technical Bulletin. d2ei442zrkqy2u.cloudfront.net/wp-content/uploads/2016/03/Zinc-Air-Tech-Bulletin.pdf.

[53] Green Car Congress, Zinc-Air Hybrid Buses Get Closer to Market, August 16, 2016. www.greencarcongress.com/2004/11/zincair_hybrid_.html.

[54] Delft University of Technology (TU Delft), Enipedia — Tesla Model S Battery. enipedia.tudelft.nl/wiki/Tesla_Model_S_Battery.

[55] A.F. Gonheim, Needs, Resources and Climate Change: Clean and Efficient Conversion Technologies, Prog. Energy Combust. Sci. 37 (2011) 15−51.

[56] S.G. Chalk, J.F. Miller, Key challenges and recent progress in batteries, fuel cells, and hydrogen storage for clean energy systems, J. Power Sources 159 (2006) 73−80. doi.org/10.1016/j.jpowsour.2006.04.058.

[57] Calculation: 75 kJ per mole of H_2 required for storage in the metal hydride divided by 285.8 kJ released in combustion per mole of H2 taking into account of the condensation of water $= 0.26 = 26\%$. If the storage penalty is 26%, then the efficiency of the storage process is 74%. The number 75 kJ per mole of H_2 is given on page 992 of .K. Zeng, T. Klassen, W. Oelerich, R. Bormann, Critical assessment and thermodynamic modeling of the Mg-H system, Int. J. Hydrogen Energy 24 (1999) 989–1004. doi.org/10.1016/S0360-3199(98) 00132-3.

[58] Calculation: 30 kJ per mole of H_2 required for storage in the metal organic framework divided by 285.8 kJ released in combustion per mole of H_2 taking into account of the condensation of water $= 0.10 = 10\%$. If the storage penalty is 10%, then the efficiency of the storage process is 90%. The number 30 kJ per mole of H_2 is given on page 16 of. P.R. Prabhukhot, M.M. Wagh, A.C. Gangal, A review on solid state hydrogen storage material, Adv. Energy Power 4 (2) 11–22. doi.org/10.13189/aep.2016.040202.

[59] A.A. Akhil, multiple other authors, DOE/EPRI 2013 Electricity Storage Handbook in Collaboration with NRECA, 2013, 164 pages + appendices, www.sandia.gov/ess/publications/SAND2013-5131.pdf.

[60] J.W. Keeney, Investigation of compressed air energy storage efficiency, M.S. thesis, in: Mechanical Engineering, California Polytechnic State University, San Luis Obispo CA, 2013, 272 pages, digitalcommons.calpoly.edu/cgi/viewcontent.cgi?article=2242&context=theses.

[61] The World Bank, World DataBank − World Development Indicators − Electric Power Transmission and Distribution Losses (% of Output). databank.worldbank.org/data//reports.aspx?source=2&country=&series=EG.ELC.LOSS.ZS&period=#.

[62] U.S. Department of Energy, Energy Information Administration − Office of Oil and Gas − Natural Gas Compressor Stations on the Interstate Pipeline Network: Developments since 1996, November 2007. www.eia.gov/pub/oil_gas/natural_gas/analysis_publications/ngcompressor/ngcompressor.pdf.

[63] C.L. Weber, H.S. Matthews, Food-miles and the relative climate impacts of food choices in the United States, Environ. Sci. Technol. 42 (2008) 3508–3513. See Table 1 based on information originating from: Davis, S. C., and S. W. Diegel, 2007: Transportation Energy Data Book: Edition 26; ORNL 6978; Oak Ridge National Laboratory, Oak Ridge, Tennessee. An updated version of this document exists: Davis, S. C., S. W. Diegel and R. G. Boundy, 2015: Transportation Energy Data Book: Edition 34; ORNL 6991; Oak Ridge National Laboratory, Oak Ridge, Tennessee, 440 pages.

[64] SPIECAPAG Entrepose, Iraq Crude Oil Pipeline Trans Saudi Arabia. www.spiecapag.com/?page=202.

[65] M.F. Ashby, Materials and the Environment, second ed., Butterworth-Heinemann, 2013, 616 pages.

[66] U.S. Energy Information Administration, How Much Electricity Does an American home Use? www.gov/tools/faqs/faq.cfm?id=97&t=3.

[67] U.S. Energy Information Administration, How Much Electricity Is Lost in Transmission and Distribution in the United States? www.eia.gov/tools/faqs/faq.cfm?id=105&t=3.

[68] University of Michigan, Center for Sustainable Systems − U.S. Food System Factsheet, Publication No. CSS01-06, October 2015 and references therein, css.snre.umich.edu/css_doc/CSS01-06.pdf.

[69] F. Jabr, Does Thinking Really Hard Burn More Calories? Scientific American, July 18, 2012, 2012. www.scientificamerican.com/article/thinking-hard-calories/.

[70] A. Marcin, How Many Calories Do You Burn while You're Asleep? Healthline, 2017, 6 November 2017 (medically reviewed), www.healthline.com/health/calories-burned-sleeping#1. This article also contains a formula for the determination of one's Basal Metabolic Rate (BMR) as a function of age, weight, height, and sex.

[71] Mayo Clinic, Weight Loss − Exercise for weight loss: Calories burned in 1 hour. www.mayoclinic.org/healthy-lifestyle/weight-loss/in-depth/exercise/art-20050999.

[72] NutriStrategy, Calories Burned during Exercise. www.nutristrategy.com/activitylist4.htm.

Chapter 4

Pollutants and greenhouse gases

4.1 Common pollutants

Air pollutants are contaminants in the atmosphere that cause direct harm to people or the environment, in contrast to greenhouse gases that affect the climate which, in turn, may cause harmful effects. Below are the names, chemical formulas (or abbreviations), and emission factors for the most common air pollutants from their most common sources. An emission factor is the ratio of the amount of pollutant emitted per amount of fuel combusted. For example, burning 1 kg of coal results in an emission of 0.3 g of CO.

Pollutant emission factors for three common fossil fuels				
		Emission factor (*EF*)		
Pollutant	**Formula**	**Coal (1.8% S) (g/kg)**	**Residual fuel oil (2% S) (g/L)**	**Natural gas (mg/m^3)**
Carbon monoxide	CO	0.3	0.6	640
Methane	CH_4	0.015	0.03	4.8
Nitrogen oxides (as NO_2 equivalent)	NO_x	10.5	8	8800
Particulate matter	"PM"	31	2.9	16–80
Sulfur dioxide	SO_2	35	38	9.6
Volatile organic compounds	"VOCs"	0.04	0.09	4.8
Source: [1] page 424.				

For additional emission factors, including other pollutants and sources, see Refs. [2,3].

Data, Statistics, and Useful Numbers for Environmental Sustainability.
https://doi.org/10.1016/B978-0-12-822958-3.00004-2

The actual emission E of an air pollutant is calculated by the following formula:

$$E = A \times EF \times \left(1 - \frac{ER}{100}\right),$$

in which A is the activity (e.g., amount of fuel combusted per time), EF is the emission factor (tabulated above), and ER is the efficiency (in %) of the emission reduction technology, if there is any ($ER = 0$ for none). For values of postcombustion emission reduction efficiencies, see Ref. [4] page 1.1−13.

The table below lists the emissions for various types of transportation. For transportation of people specifically, see Section 5.1.1 (on land), 5.2.1 (by air), and 5.3.1 (on water). For freight, see Section 5.1.2 (by road), 5.1.3 (by rail), 5.2.2 (by air), 5.3.2 (on freshwater), and 5.3.3 (on seawater).

Emissions for three types of freight transportation						
	Truck		Train		Ship	
Pollutant	g/MBTU	g/TEU-mi	g/MBTU	g/TEU-mi	g/ship-mi	g/TEU-mi
CO	3.27	1.64	213	0.39	6,869	1.37
CO_2	2,002	1,001	78,363	144.97	1,464,151	292.83
NO_x	13.73	6.86	1517	2.81	39,626	7.93
PM_{10}	0.24	0.12	36	0.07	1,173	0.23
SO_x	0.44	0.22	17	0.03	19,559	3.91
VOCs	0.68	0.34	73	0.14	1,493	0.30

MBTU, million BTUs; *TEU*, Twenty-foot Equivalent Unit, a standard container unit for shipping cargo, which is 20 ft long, 8 ft wide, and 8.5 ft high.
Source: [5].

4.2 Greenhouse gases

Excluding water vapor, which is the most potent among all greenhouse gases but with negligible anthropogenic emission, the major greenhouse gases of concern from a climate change perspective are carbon dioxide, methane, nitrous oxide, ozone, CFCs, and HCFCs. Their Global Warming Potentials defined as the equivalent amount of CO_2 (on a per-mass basis) over a 100-year horizon are given below.

Global Warming Potentials (GWPs) of various greenhouse gases			
Greenhouse gas	Formula	GWP	Atmospheric lifetime
Carbon dioxide	CO_2	1	Several centuries
Methane	CH_4	28–36	12 years
Nitrous oxide	N_2O	265–298	114 years
Tropospheric ozone[1]	O_3	–	Minutes to hours
CFC-12 (R-12)	CCl_2F_2	10,900	100 years
HCFC-22 (R-22)	$CHClF_2$	1810	12 years
Sulfur hexafluoride	SF_6	22,800	3200 years

[1]*Although it acts as a greenhouse gas, ozone is too short-lived to have a GWP.*
Sources: [6,7].

According to the Fifth Assessment Report of the Intergovernmental Panel on Climate Change [8], the various greenhouse gas emissions by economic sectors are as depicted in Figure 4-1 below.

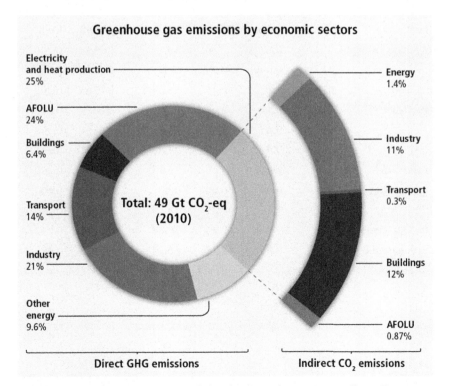

FIGURE 4-1 Global greenhouse gas emissions by economic sectors, according to Intergovernmental Panel on Climate Change (2014). *Source:* [8] Fig. 1.7.

4.3 Carbon footprints

4.3.1 Per person as function of affluence

The carbon footprint of people across the world on a per-capita basis is strongly correlated with financial expenditures. According to data analyzed in Ref. [9] for many countries, the correlation is

$$CO_{2eq} \text{ per person} = 0.0539 \text{ ($ spent per person)}^{0.555}.$$

4.3.2 Transportation fuels

The table below lists the amount of carbon dioxide of fossil origin emitted per unit amount of fuel combusted and per unit amount of energy consumed in transportation. See Chapter 5 for CO_2 emissions based on the mode of transportation and distance covered.

Carbon emissions by type of fossil fuel or biofuel						
		kg CO_2 per volume		kg CO_2 per unit of energy		
Fuel		per gallon	per liter	per 10^6 BTUs	per MJ	per kWh
Aviation gasoline		8.32	2.20	69.19	0.0656	0.236
Biodiesel	B100	0.00	0.00	0.00	0.0000	0.000
	B20	8.12	2.15	59.44	0.0563	0.203
	B10	9.13	2.41	66.35	0.0629	0.226
	B5	9.64	2.55	69.76	0.0661	0.238
	B2	9.94	2.63	71.80	0.0681	0.245
Diesel fuel		10.15	2.68	73.15	0.0693	0.250
Ethanol and blends	E100	0.00	0.00	0.00	0.0000	0.000
	E85	1.34	0.354	14.79	0.0140	0.050
Jet fuel (kerosene)		9.57	2.53	70.88	0.0672	0.242
E10 (gasohol)		8.02	2.12	66.30	0.0628	0.226

Continued

Methanol and blends	M100	4.11	1.09	63.62	0.0603	0.217
	M85	4.83	1.28	65.56	0.0621	0.224
Motor gasoline		8.91	2.35	71.26	0.0675	0.243
Propane		5.74	1.52	63.07	0.0598	0.215
Residual fuel oil (No. 5 and 6)		11.79	3.11	78.80	0.0747	0.269
Source: [10].						

4.3.3 Heating fuels

The table below lists the amount of several greenhouse gases emitted per unit amount of energy released in stationary combustion.

Greenhouse gas emissions from heating fuels							
		CO_2		CH_4		NO_x	
Fuel/Application		kg/10^6 BTUs	g/MJ	g/10^6 BTUs	mg/ MJ	g/10^6 BTUs	mg/ MJ
Coal	Commercial heating	95.35	90.37	10	9.5	1.5	1.4
	Industry	93.71	88.82	10	9.5	1.5	1.4
	Residential heating	95.35	90.37	301	286	1.5	1.4
Natural gas	Commercial	53.37	50.58	5	4.7	0.1	0.09
	Residential	53.37	50.58	5	4.7	0.1	0.09
Petroleum	Commercial	73.15	69.33	10	9.5	0.6	0.57
	Residential	73.15	69.33	10	9.5	0.6	0.57
Wood[1]	Commercial	3.7	3.5	253	240	3.2	3.0
	Residential	3.7	3.5	253	240	3.2	3.0

[1]The small CO_2 emission values for wood are attributed only to processing and transportation because the carbon in the wood itself is renewable.
Sources: [10], [11] Table 24.

Other useful numbers are as follows [10]:

- Municipal Solid Waste (MSW) generates 460 kg of CO_{2eq} per metric ton incinerated, which is equivalent to 41.70 kg of CO_{2eq} per 10^6 BTUs of heat generated;

- The plastics portion of MSW generates 2.80 kg of CO_{2eq} per kg of plastics burned;
- The burning of rubber tires generates 85.97 kg of CO_{2eq} per 10^6 BTUs generated.

4.3.4 Electricity generation

The table below lists the amounts of several greenhouse gases emitted per unit amount of electricity generated for typical sources of energy.

Carbon footprint of electricity generation			
Fuel		lb of CO_{2eq}/kWh	kg of CO_{2eq}/kWh
Coal	Bituminous	2.07	0.94
	Subbituminous	2.16	0.98
	Lignite	2.17	0.98
Fuel oil	Distillate (No. 2)	1.64	0.74
	Residual (No. 6)	1.76	0.80
Natural gas		0.92	0.42
Nuclear		0.037	0.017
Wood	Logs	<0.013	<0.0061
	Pellets	0.035	0.016
Sources: [12–14].			

The carbon footprint of electricity generation varies with the mix of fuel used, which varies from country to country. The table below gives the numbers for selected countries.

Carbon footprint of electricity generation by location		
Country	kg CO_{2eq}/kWh	Source
Brazil	0.0031	[15] p. 100
China	0.92	[16] p. 119
Europe	0.50	Ecoinvent database
Spain	0.248	[17] p. 242
United States	0.45	[12]

Renewable forms of energy are not totally carbon free when the life cycle is considered. Estimations are 14 g CO_{2eq}/kWh for geothermal, 13–731 g CO_{2eq}/kWh for solar photovoltaic, and 7–124 g CO_{2eq}/kWh for wind [13].

4.3.5 Materials

The amounts of CO_2 emitted in the production of common metals and other materials are provided in Chapter 1.

4.3.6 Buildings

Greenhouse gas emissions from buildings vary greatly. Firstly, one must distinguish the carbon emissions during use from those during construction (such as those caused in the production of concrete). Use-related emissions depend on the type of building (single family dwelling, apartment building, commercial building, *etc.*) as well as on the climate in which the building sits (necessitating various degrees of heating and/or air conditioning) and whether the building was designed with environmental considerations or not (ranging the gamut from net-zero energy buildings to centuries-old structures). Finally, there is the question of the denominator: Should the numbers be quoted per building, per square foot of floor space, or per person? Below is a gathering of rather disparate numbers gleaned from a variety of reliable sources.

The following table provides CO_{2eq} emissions from buildings per unit of end-use energy for several fuel types and two types of buildings. The overall number is 111.4 kg CO_{2eq} released per million BTUs consumed (=105.6 g CO_{2eq} per MJ).

Carbon intensity per type of fuel						
	Residential buildings		Commercial buildings		All buildings	
Fuel type	kg CO_{2eq}/ 10^6 BTUs	g CO_{2eq}/ MJ	kg CO_{2eq}/ 10^6 BTUs	g CO_{2eq}/ MJ	kg CO_{2eq}/ 10^6 BTUs	g CO_{2eq}/ MJ
Coal	95.35	90.37	95.35	90.37	95.35	90.37
Electricity	179.1	169.8	177.9	168.6	178.3	169.0
Natural gas	53.06	50.29	53.06	50.29	53.06	50.29
Petroleum	68.45	64.88	71.62	67.88	69.62	65.99
All fuels	105.6	100.1	118.7	112.5	**111.4**	**105.6**

Source: [18], based on 2010 US data.

The next table provides a breakup per end use in commercial buildings (in million metric tons of CO_{2eq} per year for the entire United States).

Carbon footprint in commercial buildings per end use						
End use	Electricity	Gas	Petroleum	Coal	Total	Percentage
Lighting	211.9	–	–	–	211.9	20.4%
Space heating	50.5	87.4	17.3	5.6	160.7	15.5%
Space cooling	149.1	2.3	–	–	151.3	14.6%
Ventilation	95.2	–	–	–	95.2	9.2%
Refrigeration	69.1	–	–	–	69.1	6.7%
Water heating	16.2	23.2	2.0	–	41.4	4.0%
Electronics	84.1	–	–	–	84.1	8.1%
Other	128.9	61.5	32.1	–	222.4	21.5%
Total	**805.0**	**174.4**	**51.3**	**5.6**	**1036.3**	**100%**

Source: [19] based on 2010 US data. "Other" includes miscellaneous activities such as cooking as well as adjustments to reconcile data from different data sources.

A typical office building in New England (Northeast USA) contributes 20 lb of CO_{2eq} per square foot (=98 kg of CO_{2eq} per m^2) of floor space, compared to an Energy Star rated building that contributes 15 lb of CO_{2eq} per square foot (=73 kg of CO_{2eq} per m^2) of floor space [20].

Figure 6-2 in Section 6.3.2 provides the annual average energy use per square foot in typical commercial buildings in the United States for 2003 and 2012. To translate these numbers into CO_{2eq} emissions, use 118.7 kg CO_{2eq} released per million BTUs consumed in commercial buildings in the United States, as provided in Ref. [21]. To obtain energy values in MJ/m^2, multiply the 1000's BTUs/ft^2 by 11.36.

According to some calculations [22], the individual CO_{2eq} footprint attributed to one's housing situation is as follows:

— Apartment (up to 1,000 ft^2): 11,000 lb (=4,990 kg) of CO_{2eq} per year;
— Small house (up to 1,500 ft^2): 16,500 lb (=7,480 kg) of CO_{2eq} per year;
— Medium house (up to 2,500 ft^2): 27,500 lb (=12,470 kg) of CO_{2eq} per year;
— Large house (up to 4,000 ft^2): 44,000 lb (=20,000 kg) of CO_{2eq} per year.

See [23] and Section 13.5 for additional carbon footprint equivalencies.

4.3.7 Food

According to a scientific study [24], the average US household's food consumption has a carbon footprint of 8.1 metric tons of CO_{2eq} per year, 11% of which is caused by various transportation activities during the life cycle including delivery of fertilizer and diesel fuel to the farm, the delivery of food to the cattle, and transportation of the harvest to a processing plant and of the food to the warehouse and store. The delivery from producer to retail store (the quantity of concern by promoters of local foods) is only 4% of the total, *i.e.*, 320 kg of CO_{2eq} per household per year. For reference, the average number of people per US household was 2.56 in 2007 [25], when the aforementioned study was conducted.

The graph below (Figure 4-2) breaks down the average US household's carbon footprint per food type, with red meat topping the chart by a wide margin.

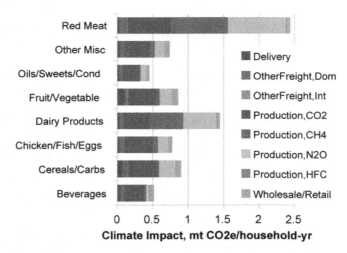

FIGURE 4-2   Carbon emissions in food production, processing, and transportation, in metric tons of CO_{2eq} per household per year. *Source:* [24] Fig. 1c.

The table below provides the carbon footprint of a variety of food items. As a rule, naturally grown fruits and vegetables have no initial footprint unless the ground around the plant is fertilized or irrigated, and most of the ultimate footprint is a result of handling such as bottling, cold storage, and transportation. Some additional impacts arise from cooking.

Carbon footprint of various foods			
Food item/Handling		CO_{2eq} release[1]	Remark
Apples	Plucked from one's garden	0.0	
	Local and seasonal	10 g	
	Average	80 g	=550 g per kg
	Shipped and cold stored	150 g	
Asparagus	Local and seasonal	125 g	For a 250 g bunch
	Air-freighted across hemispheres	1.9 kg	
	Air-freighted halfway across world	3.5 kg	
Bananas	Grown in one's garden	0.0	
	Imported from across the world	80 g	=480 g per kg
Beer	Locally brewed, on tap	300g	Per pint
	Locally brewed and bottled	500 g	
	Transported, on tap	500 g	
	Bottled and transported	900 g	
Bread		1 kg	For an 800 g loaf
Carrots	Local and seasonal	0.25 kg	Per kg
	Average	0.3 kg	
	Shipped baby carrots	1 kg	
Coffee	Boiling just enough water	23 g	For one serving
	Same, with milk	55 g	
	Large cappuccino	236 g	
	Large latte	343 g	
Dairy	Ice cream, consumed on day of purchase	50 g	Per serving
	Cheese	12 kg	Per kg
	Milk	723 g	Per pint

Continued

Eggs	Raw	3.6 kg	For a dozen
Meats	Bacon	200 g	For a 25-g slice
	Beef	2 kg	For a raw 4-oz steak
	Cheeseburger	2.5 kg	For 4-oz burger
	Lamb, leg at store	38 kg	For one leg
Oranges	Shipped 2000 miles	90 g	=500 g per kg
Porridge	Traditional Scottish	82 g	Per bowl
Potatoes	Local, boiled with lid on	620 g	Per kg
Rice	Average	4 kg	Per kg
Seafood	Fresh, landed	0.5−2.6 kg	Per kg
	Fresh, at store	2.8−3.2 kg	
	Frozen	3.2−6.5 kg	
	Shrimp, peeled, and frozen	10.1 kg	
Strawberries	Local and seasonal	150 g	Small basket
	Imported or out of season	1.8 kg	
Tomatoes	Organic, local, traditional	0.4 kg	Per kg
	Average	9.1 kg	
	Cherry tomatoes, transported	50 kg	
Wine	Growing grapes	0.28−0.85 kg	Per kg
	Average type and transportation	1.04 kg	Per bottle

[1]Unless otherwise noted, the quoted number is per piece (e.g., one apple or one banana).
Sources: [26,27].

4.4 Carbon sequestration and offsets

4.4.1 Trees and other biomass

According to Ref. [28], the amount of carbon dioxide that a tree sequesters during its growth ranges from 2.5 to 14 lb (1.14−6.4 kg) per year per tree depending on the type of tree and climatic region.

			Rate of CO_2 sequestration		
Forest type	Age (years)	lb/yr/ tree	kg/yr/ tree	lb/yr/ acre[1]	kg/yr/ hectare[1]
Maple–beech– birch forest	25	2.52	1.14	1,760	1,970
Maple–beech– birch forest	120	5.58	2.53	3,909	4,380
White and red pine forest	25	14.0	6.37	9,826	11,010
White and red pine forest	120	10.7	4.87	7,516	8,420

[1]*Based on 700 trees per acre = 1730 trees per hectare.*
Source: [28] based on data from the Northeast USA.

The average of the preceding numbers is 8.20 lbs per year per tree (3.72 kg per year per tree). Thus, a 1-acre (= 4,047 m^2 = 0.4047 ha) parcel of trees planted for a sequestration project will sequester over a 70-year lifetime 402,000 lbs (=182 metric tons) of CO_2. This number is comparable to the one quoted in Ref. [28] that mentions a 410-acre parcel sequestering 70,000 short tons (=63,500 metric tons) over 70 years. Figure 4-3 below shows the ability of common trees to absorb carbon over time.

The table below provides a comparison of forests with cropland, grassland, and wetlands. Beware of the wide range of values.

Carbon sequestration rates for various types of vegetation		
	Range of CO_2 sequestration rate	
Biome	(metric tons/acre/year)	(metric tons/ha/year)
Cropland	0.2–0.6	0.5–1.5
Forest	0.05–3.9	0.12–9.6
US timberland	0.21–0.23	0.51–0.56
Grassland	0.12–1.0	0.30–2.5
Swamp/floodplain/wetland	2.23–3.71	5.51–9.17

Sources: [29,30].

According to Ref. [29], world forest vegetation and soils contain an estimated 1146 billion metric tons of carbon (elemental C) with approximately 37% in low latitudes, 14% in midlatitudes, and 49% at high latitudes. For the United States, the carbon content is 71 billion metric tons of carbon distributed over 302.3 million hectares (33% of land), and the carbon uptake rate ranges between 155 and 170 metric tons of C per year.

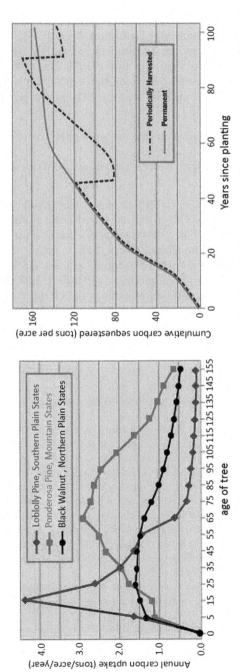

FIGURE 4-3 Carbon sequestration rates for trees in the United States. *Left panel*: Elemental carbon uptake over the lifetime of common trees. *Right panel*: Carbon accumulation rate for loblolly pine. *Source*: [29] Fig. 7-1.

4.4.2 Underground injection

According to the US Environmental Protection Agency [31], carbon capture and sequestration technologies can dramatically reduce (by 80–90%) the CO_2 emissions from power plants that burn fossil fuels. For example, if 90% of greenhouse gas emissions could be avoided at a 500 MW coal-fired power plant that emits roughly 3 million tons of CO_2 per year, this would be equivalent to

- Planting more than 62 million trees, and waiting at least 10 years for them to grow;
- Avoiding annual electricity-related emissions from more than 300,000 homes.

The US Department of Energy [30] estimates that anywhere from 1800 to 20,000 billion metric tons of CO_2 could be stored underground in the United States. This is equivalent to 600 to 6700 years of current level emissions from large stationary sources in the United States [31].

Carbon sequestration by in-ground injection requires 4.6 kg of CO_2 to be injected for every 1 kg of crude oil produced (15,000 ft^3 of CO_2 per barrel of crude oil, at standard temperature and pressure) [32]. Of this CO_2 injected, half is assumed to be produced with the crude oil extracted, which is subsequently removed and reinjected. The remaining 2.3 kg of CO_2 gas (per kg of crude oil extracted) is manufactured and injected.

Sources

[1] W.W. Nazaroff, L. Alvarez-Cohen, Environmental Engineering and Science, John Wiley & Sons, 2001, 690 pages.

[2] U.S. Environmental protection Agency − Technology Transfer Network − Clearinghouse for Inventories and & Emission Factors. www3.epa.gov/ttnchie1/ap42/.

[3] SYKE − Air Pollutant Emission Factor Library. www.apef-library.fi/.

[4] U.S. Environmental Protection Agency − External Combustion Sources − Bituminous and Subbituminous Coal Combustion. www3.epa.gov/ttn/chief/ap42/ch01/final/c01s01.pdf.

[5] J.J. Corbett, J.J. Winebrake, J. Hatcher, and A.E. Farrell, reportundated (circa 2007): Emissions Analysis of Freight Transport Comparing Land-Side and Water-Side Short-Sea Routes: Development and Demonstration of a Freight Routing and Emissions Analysis Tool (FREAT), Final Report for the Research and Special Programs Administration at the U.S. Department of Transportation, 45 pages. www.transportation.gov/sites/dot.gov/files/docs/emissions_analysis_of_freight.pdf.

[6] U.S. Environmental Protection Agency − Greenhouse Gas Emissions − Understanding Global Warming Potentials. www.epa.gov/ghgemissions/understanding-global-warming-potentials.

[7] Center for Climate and Energy Solutions (C2ES) − Main Greenhouse Gases. www.c2es.org/facts-figures/main-ghgs.

[8] Intergovernmental Panel on Climate Change (IPCC) − Fifth Assessment Report − AR5 Synthesis Report: Climate Change 2014. www.ipcc.ch/report/ar5/syr/synthesis-report/.

[9]   E.G. Hertwich, G.P. Peters, Carbon footprint of nations: a global, trade-linked analysis, Environ. Sci. Technol. 43 (2009) 6414—6420, doi.org/10.1021/es803496a.

[10] U.S. Energy Information Administration (eia) — Environment.
   1. Carbon Emission Factors for Stationary Combustion
   2. Carbon Dioxide Emission Factors for Transportation Fuels
   3. Generic Methane and Nitrogen Oxide Emission Factors for Stationary Fuel Combustion. www.eia.gov/oiaf/1605/coefficients.html

[11] Biomass Energy Resource Center (Burlington, Vermont, USA) — Biomass Supply and Carbon Accounting for Southeastern Forests — February 2012, 132 pages. www.biomasscenter.org/images/stories/SE_Carbon_Study_FINAL_2-6-12.pdf.

[12] U.S. Energy Information Administration (EIA) — Independent Statistics & Analysis — Frequently Asked Questions: How much carbon dioxide is produced per kilowatthour when generating electricity with fossil fuels? www.eia.gov/tools/faqs/faq.cfm?id=74&t=11.

[13] Nuclear Energy Institute — Issues & Policy — Protecting the Environment — Life-Cycle Emissions Analyses. www.nei.org/Issues-Policy/Protecting-the-Environment/Life-Cycle-Emissions-Analyses.

[14] Stoves Online — CO_2 Heating Fuel Emissions. www.stovesonline.co.uk/fuel-CO2-emissions.html.

[15] L.S. Franca, M.S.R. Rocha, G.M. Ribeiro, Carbon footprint of municipal solid waste considering selective collection of recyclable waste. Chap. 4, in: S.S. Muthu (Ed.), Environmental Carbon Footprints — Industrial Case Studies, Butterworth-Heinemann, Elsevier, 2018, pp. 79—112, doi.org/10.1016/B978-0-12-812849-7.00004-0.

[16] W.K.C. Yung, S.S. Muthu, K. Subramanian, Carbon footprint analysis of personal electronic product — induction cooker. Chap. 5 in: S.S. Muthu (Ed.), Environmental Carbon Footprints — Industrial Case Studies, Butterworth-Heinemann, Elsevier, 2018, pp. 113—140, doi.org/10.1016/B978-0-12-812849-7.00005-2.

[17] A. Martínez-Rocamora, J. Solís-Guzmán, M. Marrero, Carbon footprint of utility consumption and cleaning tasks in buildings. Chap. 9 in: S.S. Muthu (Ed.), Environmental Carbon Footprints — Industrial Case Studies, Butterworth-Heinemann, Elsevier, 2018, pp. 229—258, doi.org/10.1016/B978-0-12-812849-7.00009-X.

[18] U.S. Department of Energy — Energy Efficiency & Renewable Energy — Buildings Energy Data Book — 2010 Carbon Dioxide Emission Coefficients for Buildings. buildingsdatabook.eren.doe.gov/TableView.aspx?table=1.4.8.

[19] U.S. Department of Energy — Energy Efficiency & Renewable Energy — Buildings Energy Data Book — 2010 Commercial Buildings Energy End-Use Carbon Dioxide Emissions Splits, by Fuel Type. buildingsdatabook.eren.doe.gov/TableView.aspx?table=3.4.2.

[20] U.S. Environmental Protection Agency — Newsroom — New Tool Offers and Inside Look at the Climate Change Impact of Buildings (dated 28 Sept 2007). yosemite.epa.gov/opa/admpress.nsf/6427a6b7538955c585257359003f0230/44d2fe4a462008f38525736400519e43!OpenDocument.

[21] U.S. Energy Information Administration (eia) — Commercial Buildings Energy Consumption Survey (CBES) — 2012 Commercial Building Energy Consumption Survey: Energy Usage Summary. www.eia.gov/consumption/commercial/reports/2012/energyusage/index.php.

[22] Carbonfund.org — Reduce Your Individual Carbon Footprint. carbonfund.org/individuals/.

[23] U.S. Department of Energy — Energy Efficiency & Renewable Energy — Buildings Energy Data Book — 1.5: Generic Fuel Quad and Comparison — 1.5.3 Carbon Emissions Comparisons. buildingsdatabook.eren.doe.gov/TableView.aspx?table=1.5.3.

[24] C.L. Weber, H.S. Matthews, Food-miles and the relative climate impacts of food choices in the United States, Environ. Sci. Technol. 42 (2008) 3508—3513, doi.org/10.1021/es702969f.

[25] Statista — Demographics - Number of People per Household in the United States from 1960 to 2015. www.statista.com/statistics/183648/average-size-of-households-in-the-us/.

[26] M. Berners-Lee, How Bad Are Bananas? The Carbon Footprint of Everything, Greystone Books, D&M Publishers, 2011, 232 pages.

[27] V.D. Litskas, T. Irakleous, N. Tzortzakis, M.C. Stavrinides, Determining the carbon footprint of indigenous and introduced grape varieties through Life Cycle Assessment using the island of Cyprus as a case study, J. Clean. Prod. 156 (2017) 418—425.

[28] Tufts University — Office of Sustainability — Carbon Sequestration. sustainability.tufts.edu/carbon-sequestration/.

[29] R.W. Malmsheimer, P. Heffernan, S. Brink, D. Crandall, F. Deneke, C. Galik, E. Gee, J.A. Helms, N. McClure, M. Mortimer, S. Ruddell, M. Smith, J. Stewart, Forest management solutions for mitigating climate change in the United States, J. Forestry 106 (3) (2008) 115—173, with a 2-page correction published in the same journal in June 2009. In particular: Chapter 7 — Reducing Atmospheric GHGs through Sequestration, pages 148—156.

[30] Tribal Energy and Environmental Information (TEEIC) — Office of Indian Energy and Economic Development — Terrestrial Sequestration of Carbon Dioxide. teeic.indianaffairs. gov/er/carbon/apptech/terrapp/index.htm.

[31] U.S. Environmental Protection Agency — Climate Change — Carbon Dioxide Capture and Sequestration. www3.epa.gov/climatechange/c.

[32] Piedmont Biofuels — Life Cycle — Chapter 4: Petroleum Diesel Fuel Modeling. www. biofuels.coop/archive/lifecycle_ch4.pdf.

Chapter 5

Transportation

The transportation sector contributes 15% to the world's greenhouse gas emissions from fossil fuels [1]. In the United States, the share is 29% [2]. The near totality of greenhouse gas emissions from transport operations is in the form of CO_2 emissions from the direct or indirect combustion of fossil fuels.

Various modes of transportation by land, air, and water use different types of fuel sources as shown in (Figure 5-1).

Per passenger or per unit of freight, different modes of transportation have vastly different energy requirements and carbon footprints. For the transport of freight, the metric is energy (or CO_{2eq}) per unit ton and per km, whereas for the transport of people it is per passenger and per km. Helpful unit conversion factors are the following:

1.0 metric ton × km = 1,370 lbs × mile;
1.0 MJ per km = 1,525 BTUs per mile;
1.0 MJ per metric ton and per km = 1,383 BTUs per short ton and per mile;
1 hp = 0.7457 kW;
(# mpg) × (# L/100 km) = 235.

5.1 Land transport

5.1.1 Land transport of people

A midsize car weighs 1,590 kg (3,500 lbs) while an SUV or compact truck weighs 1,577 kg (3,470 lbs) [4], and both nominally carry four passengers. Their average fuel efficiency in the United States is 24.9 mpg (miles per gallon), equivalent to 9.45 L/100 km, and their CO_2 emission is estimated at 357 grams per mile (222 g/km) [5]. For a hybrid car with a fuel consumption of 45 mpg (81% higher than a conventional car), the carbon footprint drops to 198 g/mi (123 g/km).

In the United States, the automobile is driven an average of 12,000 miles (19,300 km) per adult per year and requires 170 hp (127 kW) of power when driven at 55 mph (=89 km/h) [6].

Data, Statistics, and Useful Numbers for Environmental Sustainability.
https://doi.org/10.1016/B978-0-12-822958-3.00014-5

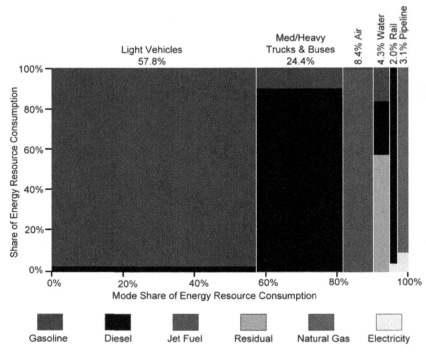

FIGURE 5-1 Consumption of energy for transportation in the United States, by mode and fuel type. Courtesy of Oak Ridge National Laboratory, U.S. Dept. of Energy. *Source:* [3] Fig. 2.6.

From these numbers, one obtains the conversion of tons of CO_{2eq} into cars driven, which can be used to convert tons of CO_2 avoided into an equivalent number of cars taken off the road:

$$1 \, \frac{\text{metric ton } CO_2}{\text{year}} \, \text{saved} \rightarrow \frac{(1 \text{ ton } CO_2/\text{yr})(10^6 \text{g}/\text{ton})}{(357 \text{ g}/\text{mi})(12{,}000 \text{ mi}/\text{car} \times \text{yr})}$$
$$= 0.233 \text{ cars off the road.}$$

The tables below provide the energy required and CO_{2eq} emissions to transport one passenger for a unit distance depending on the mode of transportation, in Europe and in the United States.

Passenger transportation in Europe				
		Energy consumption		CO$_2$ emission
Mode of transportation in Europe		BTUs/ passenger-mile	MJ/ passenger-km	lb CO$_2$/ passenger-mile	g CO$_2$/ passenger-km
Automobile	diesel	1,800 –2,750	1.2–1.8	0.493	139
	gasoline			0.539	152

Continued

SUV, light truck		2,420 −3,630	1.6−2.4	0.713	201
Motorcycle (motorcyclist only)		1,800	1.2	0.414	117
Bus	urban, 10 passengers	900	0.6	0.426	120
	urban, >20 passengers	600	0.40	0.289	81.6
Long-distance motor coach		290	0.19	0.0968	27.3
Light rail tram		490	0.32	0.0976	27.5
Subway/metro		730	0.48	0.146	41.2

Sources: [7] Table 1, [8], [9] Tables 20, 24−26 with some averages from multiple values. Some energy values prorated from known CO_2 emissions for transportation modes with same fuel.

Passenger transportation in the United States				
	Energy consumption		CO_2 emission	
Mode of transportation in the United States	BTUs/ passenger- mile	MJ/ passenger- km	lb CO_2/ passenger- mile	g CO_2/ passenger- km
Automobile (gasoline)	2,890 −3,585	1.90−2.35	0.82	231
SUV, light truck	3,390 −4,233	2.23−2.78	1.00	281
Motorcycle (motorcyclist only)	2,400 −2,450	1.58−1.61	0.57	159
Urban transit bus[1]	3,319 −4,535	2.18−2.98	0.66	186
Highway motor coach	749−804	0.49−0.53	0.123	34.8
Subway	1,210	0.79	0.27 −0.315	76−89

[1]*High value because ridership rate is only on the order of 25% capacity and the vehicle makes frequent stops and idles in dense traffic.*
Sources: [3] Tables 2.13 and 2.14, [10] Table 4-20, [11] Table 1.1, [12,13]. Some CO_2 emissions prorated from known energy values using ratios from preceding table.

The profile of several types of passenger trains is shown below.

	Passenger transportation by trains			
	Energy consumption		CO_2 emission	
Train type	10^6 BTUs/ passenger-mile	MJ/ passenger-km	lb CO_2/ passenger-mile	g CO_2/ passenger-km
UK/EU rail	1,260–1,540	0.83–1.01	0.21	60
US transit rail	788–1,608	0.52–1.05	0.39	110
US intercity train	1,607–2,091	1.05–1.37	0.60	170
AMTRAK, intercity	1,524	1.00	–	–
US electric train	–	–	1.21	340

Sources: [3] Table 2.15, [7] Table 1, [9] Table 26, [10] Table 4-20, [11] Table 1.1.

5.1.2 Land transport of freight by road

In the United States, a fully loaded 18-wheel diesel freight truck has a fuel efficiency of 5.9 mpg (=40 L/100 km) ([3] Table 5.2, [14]), rising to 8.0 mpg (=29.2 L/100 km) on open European roads [15].

The table below provides the energy required and carbon footprint to transport a unit of freight over a unit distance depending on the mode of transportation.

Freight transportation on roadways		
Mode of transportation	Energy (MJ/ton-km)	Carbon emissions (g CO_2/ton-km)
Tractor trailer diesel truck (up to 55 tons)	0.71	50
Diesel truck — up to 40 tons	0.82	60
Diesel truck — up to 32 tons	0.94	67
Diesel truck — up to 14 tons	1.5	110
Light goods vehicle, delivery van	2.5	180

Continued

Family car	Diesel	1.4–2.0	100–140
	Gasoline	2.2–3.0	140–190
	Hybrid	1.55	100
Sports utility vehicle (SUV)		4.8	310

Sources: [15] Fig. 6, [16] Table 6.9, [17] Tables 2, 5 & 8.

5.1.3 Land transport of freight by rail

Rail transportation is a more environmentally friendly way than road transportation to move freight over land. The rule of thumb is that a train can move a (short) ton of freight more than 470 miles on 1 gallon of diesel fuel [18]. This amounts to 1 L of fuel hauling 1 metric ton over 180 km (or 180 metric tons over 1 km).

The energy intensity of rail transportation in the United States is 0.19 MJ/ton-km ([3] Table 10.8), and its carbon footprint is 18 g of CO_{2e}/ton-km ([19] Table 1), whereas in Europe it is 0.33 MJ/ton-km and 25 g of CO_{2e}/ton-km [20]. After the inclusion of the energy required to process and deliver the diesel fuel and to power equipment and buildings, train freight demands 0.43 MJ of primary energy per ton-km [15,19], about a third more than the traction energy.

Intermodal operations have different values for emissions, with the table below showing a range of emission factors for different types of intermodal service with the road share of the total distance traveled varying from 5 to 20%.

Intermodal freight transportation on land				
	gCO_2/(ton-km) function of road as % of total travel distance			
	5%	10%	15%	20%
Road—rail	24	26	28	30
Road—inland waterway	32.6	34.1	35.7	37.2
Road—short-sea	18.3	20.6	22.9	25.2

Source: [17] Table 6.

5.2 Air transport

Jet engines emit a range of chemicals during operation; approximately 70% of the total emitted mass is CO_2 and 30% is water. Less than 1% of the mass consists of carbon monoxide (CO), unburned hydrocarbons (HCs, which include volatile organic compounds or VOCs), nitrogen oxides (NO_x), sulfur oxides (SO_x), particle matter (PM), and trace compounds like metals [21].

The aviation industry consumes around 1.5 billion barrels of Jet A-1 fuel annually [22], a number that varies with the state of the world economy. Worldwide in 2018, aviation consumed 359 billion L (95 billion gallons) of jet fuel [23], producing 895 million tons of CO_2, to transport 4.4 billion passengers [22], resulting in 81.6 L of jet fuel and 203 kg of CO_2 per passenger. The global aviation industry contributes to about 2% of all human-induced carbon emissions or 12% from all transport sources [22].

Jet aircrafts in 2018 were 80% more fuel-efficient than their predecessors in the 1960s [22]. The United States airline fuel efficiency in 2008 was approximately 24.6 km/L ([24] page 18) with aircraft consumption in the range of 0.0465 L of jet fuel per passenger per kilometer on flights longer than 5,000 km [22,24].

5.2.1 Passenger air travel

Energy consumption per passenger and per unit distance varies with the type of aircraft as tabulated below. Numbers have decreased over the years as jet engines have become more efficient, aircrafts hold more seats, and online bookings ensure full planes. In 2017, aircrafts flown at full occupancy were only slightly more energy consumptive than intercity trains running at partial occupancy.

Passenger transportation by air		
Type of aircraft	BTUs/passenger-mile	MJ/passenger-km
Private airplane	9,600–11,100	6.3–7.3
Air carriers—older	6,700	4.4
Air carriers—2017	2,415	1.59

Sources: [3] Table 2.15, [7] Table 1, [10] Table 4-20.

Aviation uses two types of liquid fuel, jet kerosene for jet engines and aviation gasoline for piston-powered airplanes. Jet kerosene is the most common. These fuels have the following carbon emissions per volume combusted.

Carbon emissions of aviation fuels		
Liquid fuel	kg of CO_2/gallon	kg of CO_2/L
Jet kerosene	10.15	2.68
Aviation gasoline "Avgas"	8.31	2.20

Source: [25].

Combining the fuel consumption with the CO_2 emission per volume of fuel consumed, we obtain (2.68 kg CO_2/L) × (1.59 MJ/passenger-km)/(37.6 MJ/L) = 113 g of CO_2 per passenger per kilometer (0.402 lbs of CO_2 per passenger per mile).

Because there is only one takeoff and one landing per flight regardless of the distance covered, aviation statistics are generally grouped into three categories, for which the carbon emissions per passenger per unit distance differ, as tabulated in the following.

Carbon footprint of passenger flights				
Distance			g of CO_2/ passenger-km	lbs of CO_2/ passenger-mile
Short haul	<452 km	(<337 mi)	123.9	0.44
Medium haul	452 −1,600 km	(337 −1,000 mi)	78.9	0.28
Long haul	>1,600 km	(>1,000 mi)	111.5	0.40
Source: [9] Table 36.				

5.2.2 Airfreight

The following table lists the estimates of the energy consumption and carbon footprint depending on distance traveled and type of aircraft.

Freight transportation by air				
	Energy consumption		Carbon footprint	
Type of flight	BTUs /ton-mile	MJ /ton-km	lbs of CO_2 /ton-mile	kg of CO_2 /ton-km
Short haul	17,000−23,000	11−15	2.7−3.5	0.76−1.0
Medium haul	9,900−17,000	6.5−11	1.6−2.7	0.45−0.76
Long haul	9,900	6.5	1.4−1.6	0.40−0.45
Average inside UK	56,000	37	8.9	2.5
Average inside US	15,000	10.0	2.4	0.68
Helicopter	84,000	55	12	3.30
Sources: [9] Table 39, [16] Table 6.9, [18] Table 1.				

5.3 Water transport

With the exception of ferry boats, which are few, transportation on the water carries freight, not passengers.

5.3.1 Passengers on ferry boat

A ferry boat consumes about 11,000 BTUs per passenger-mile (=7.2 MJ per passenger-km) and emits 820 g of CO_2 per passenger per mile ([11] Table 1.1).

5.3.2 Freight on inland water

On average, boats on inland waterways consume 0.23 MJ/ton-km (=350 BTUs/ton-mi) when the cargo is in bulk and 0.22 MJ/ton-km (=336 BTUs/ton-mi) when the cargo consists of containers ([15] Fig. 6).

5.3.3 Freight on seawater

On average, ships carrying bulk cargo across the ocean consume 0.2 MJ/ton-km (=300 BTUs/ton-mi) and emit 11 g CO_2/ton-km (=0.039 lbs CO_2/ton-mi) whereas ships carrying cargo in containers consume 0.2 MJ/ton-km (=300 BTUs/ton-mi) and emit 14 g CO_2/ton-km (=0.050 lbs CO_2/ton-mi) ([17] Table 5, [19] Table 1).

Oil tankers are more efficient, consuming only 0.1 MJ/ton-km (=150 BTUs/ton-mi) and emitting approximately 5 to 7 g CO_2/ton-km (=0.018 to 0.025 lbs CO_2/ton-mi) ([1] Table 5, [19] Table 1).

5.4 Pipelines

See also Section 3.6.2.

Pipelines offer an alternative method for the transport of certain liquids (such as oil, ethylene, and propylene) and gases (such as natural gas). To overcome frictional losses along the way, an initial compression is required with possibly additional intermediate stages of compression at locations along the way. The electrical energy required to power the compressors is easily metered and therefore known.

A 2003 European report [26] estimates the energy demand at 0.046 MJ/ton-km for the compression of a liquid, yielding an estimate of 0.12 MJ/ton-km in terms of primary energy (assuming a 38% of energy loss in the generation and transmission of the electricity). This same report provides a bracket of 0.11−0.18 MJ/ton-km, with an average of 0.14 MJ/ton-km of primary energy for the sector, albeit from an older data set.

For pipeline transport of ethylene and propylene, specifically, primary energy estimates are 0.13 MJ/ton-km and 0.12 MJ/ton-km, respectively [26]. The carbon footprint of conveying liquid chemicals by pipeline is pegged at 5 g CO_{2eq}/ton-km ([17] Table 8).

According to Table 1 in Ref. [19], oil pipelines necessitate 0.2 MJ/ton-km (=300 BTUs/ton-mile) causing a carbon footprint of 16 g CO_{2eq}/ton-km, whereas gas pipelines necessitate 1.7 MJ/ton-km (=2,600 BTUs/ton-mile) causing a carbon footprint of 180 g CO_{2eq}/ton-km. Using the latter data, we can infer a carbon footprint of about 93 g CO_{2eq} per MJ (=26 g CO_{2eq} per kWh) of primary energy, on the low side given numbers quoted in Section 4.3.4 for the carbon footprint of electricity generation.

5.5 Electric cars

It is difficult to estimate the carbon footprint of an electric car, for it strongly depends on the local electricity generation mix. While the focus is on the cleaner use phase of the vehicle, a study has estimated that electric vehicles reduce global warming potential by 20—24% compared to gasoline internal combustion engine vehicles and by 10—14% relative to diesel ICEVs under the assumption of a 150,000 km vehicle lifetime unless the electricity is generated from the combustion of coal (as in China) in which case electric cars cause an increase in global warming potential of 17—27% compared with conventional cars [27].

With a 22% reduction in global warming potential compared to a midsize car that emits 222 g CO_{2eq} per km [5] (Section 5.1.1), the electric car is responsible for 173 g CO_{2eq} per km.

The human toxicity potential over the life cycle of electric cars is also estimated to be 180—290% higher than for cars with internal combustion engines, due to their significantly higher amount of copper wiring and combination of lithium, nickel, cobalt, and manganese in their batteries [27].

5.6 Bicycling versus driving

Bicycling is often touted as a green alternative to driving, but how do the numbers actually compare? Surely, a bicycle has no tailpipe, but the bicyclist's pedaling power, estimated at 163 W [28], comes from eating food, and food has an average carbon footprint of 4.4 g CO_{2eq} per kcal (1.05 mg CO_{2eq}/J) [19]. In addition, the metabolic efficiency of our human bodies to process food into mechanical energy is not very high, about 12.5% at a speed of 20 km/h (12 mph) for the average individual and rising to 20% at 30 km/h (19 mph) [28].

Thus, at a speed of 25 km/h (15.5 mph) with intermediate metabolic efficiency of 16.3%, the bicyclist has a carbon footprint of 151 g CO_{2eq} per km. By comparison, a gas-powered midsize car generates 222 g CO_{2eq} per km ([5] and Section 5.1.1). Thus, bicycling is the cleaner option if the driver is alone in the car. With one passenger, the car's emission falls to 111 g CO_{2eq}/km per passenger, below that of two bicyclists each responsible for151 g CO_{2eq}/km. A hybrid car with a fuel efficiency of 45 mpg (5.2 L/100 km) emits 123 g CO_{2eq}/km and beats the bicyclist even with only the driver on board. An electric car recharged by electricity from the grid has, in average, a carbon emission of 173 g CO_{2eq} per km (Section 5.5), making it slightly less preferable to bicycling.

Bicycling has indirect benefits over driving, not the least of which are health benefits and the much lower need for parking spaces, each of which is difficult to quantify.

Sources

[1] Center for Climate and Energy Solutions (C2ES). Climate Basics — Energy/Emissions Data — Global Emissions. www.c2es.org/content/international-emissions/.

[2] U.S. Environmental Protection Agency. Greenhouse Gas Emissions — Sources of Greenhouse Gas Emissions — Overview. www.epa.gov/ghgemissions/sources-greenhouse-gas-emissions.

[3] S.C. Davis, R.G. Boundy, Transportation Energy Data Book: Edition 38, Oak Ridge National Laboratory, U.S. Dept. of Energy, 2020, 449 pages, tedb.ornl.gov/wp-content/uploads/2020/02/TEDB_Ed_38.pdf.

[4] lovetoknow.com. List of Car Weights, by Kate Miller-Wilson, undated. cars.lovetoknow.com/List_of_Car_Weights.

[5] U.S. Environmental Protection Agency (EPA). Highlights of the Automotive Trends Report — Figure ES-1. Updated 6 March 2019. www.epa.gov/automotive-trends/highlights-automotive-trends-report.

[6] U.S. Energy information Administration (eia), Household Vehicles Energy Use: Latest Data & Trends, November 2005. www.eia.gov/consumption/residential/pdf/046405.pdf.

[7] P.J. Pérez-Martínez, I.A. Sorba, Energy Consumption of passenger land transport modes, Energy Environ. 21 (6) (2010) 577—600, doi.org/10.1260/0958-305X.21.6.577.

[8] The Odyssee-Mure Project — Sectoral Profile — Transport — Energy consumption. www.odyssee-mure.eu/publications/efficiency-by-sector/transport/transport-eu.pdf.

[9] U.K. Department for Business, Energy & Industrial Strategy — 2019 Government Greenhouse Gas Conversion Factors for Company Reporting, August 2019, 129 pages, assets.publishing.service.gov.uk/government/uploads/system/uploads/attachment_data/file/829336/2019_Green-house-gas-reporting-methodology.pdf.

[10] U.S. Department of Transportation (DoT). Bureau of Transportation Statistics — Energy Intensity of Passenger Modes. www.bts.gov/content/energy-intensity-passenger-modes.

[11] Comparison of Energy Use & CO_2 Emissions from Different Transportation Modes, Report prepared by M. J. Bradley & Associates for the American Bus Association, May 2007, 17 pages, jamesrivertrans.com/wp-content/uploads/2012/05/ComparativeEnergy.pdf.

[12] W. Brownsberger, Massachusetts State Senator, Transit Energy Efficiency, August 18 , 2019. 8-page release, willbrownsberger.com/transit-energy-efficiency/.

[13] Metropolitan Transportation Authority — MTA Press Releases — MTA Reduces Carbon Emissions and Moves Forward with Earth-Friendly Efforts. www.mta.info/press-release/mta-headquarters/mta-reduces-carbon-emissions-and-moves-forward-earth-friendly-efforts.

[14] T. Tuft, How to Improve Fuel Efficiency on the Road, *Truckinginfo*, Heavy Duty Trucking, internet posting, 4 June 2013, 2013. www.truckinginfo.com/152861/how-to-improve-fuel-efficiency-on-the-road.

[15] PLANCO Consulting GmbH, Essen — Economical and Ecological Comparison of Transport Modes: Road, Railways, Inland Waterways — Summary of Findings, November 2007. www.ebu-uenf.org/fileupload/SummaryStudy_engl.pdf.

[16] M.F. Ashby, Materials and the Environment — Eco-Informed Material Choice, second ed., Butterworth-Heinemann, 2013, 616 pages.

[17] The European Chemical Industry Council (CEFIC), Measuring and Managing CO2 Emissions of European Chemical Transport — Report authored by Alan McKinnon and Maja Piecyk, Logistics Research Centre, Heriot-Watt University, Edinburgh UK, January 24, 2011, 40 pages, cefic.org/app/uploads/2018/12/MeasuringAndManagingCO2EmissionOfEuropean Transport-McKinnon-24.01.2011-REPORT_TRANSPORT_AND_LOGISTICS.pdf.

[18] Association of American Railroads. Freight Railroads Embrace Sustainability & Environmental Preservation — Factsheet. www.aar.org/wp-content/uploads/2019/02/AAR-Sustainability-Fact-Sheet-2019.pdf.

[19] C.L. Weber, H.S. Matthews, Food-miles and the relative climate impacts of food choices in the United States, Environ. Sci. Technol. 42 (2008) 3508—3513, doi.org/10.1021/es702969f.

[20] Union Internationale des Chemins de Fer (International Union of Railways), Rail Transport and Environment — Facts & Figures, June 2008, 40 pages, siteresources.worldbank.org/ EXTRAILWAYS/Resources/515244-1268663980770/environment.pdf.

[21] J Christopher, Sequeira: Relationships between Emissions-Related Aviation Regulations and Human Health, Dept. of Aeronautics & Astronautics, MIT, 2008, 22 pages. web.mit. edu/aeroastro/partner/reports/hartman/sequeira-08.pdf. For further information, see:Gayle Ratliff, Christopher Sequeira, Ian Waitz, Melissa Ohsfeldt, Theordore Thrasher, Michael Graham and Terence Thompson: Aircraft Impacts on Local and Regional Air Quality in the United States, PARTNER Project 15 final report, October 2009, 181 pages. web.mit.edu/ aeroastro/partner/reports/proj15/proj15finalreport.pdf.

[22] Air Transport Action Group (ATAG). Facts & Figures. www.atag.org/facts-figures.html.

[23] T. Statista, A. Logistics, Total fuel consumption of commercial airlines worldwide between 2005 and 2020. www.statista.com/statistics/655057/fuel-consumption-of-airlines-worldwide/.

[24] C. Laszlo, K. Christensen, D. Fogel, G. Wagner, P. Whitehouse (Eds.), Berkshire Encyclopedia of Sustainability, vol. 2, The Business of Sustainability, 2009, 530 pages, with specific reference to Airline Industry on pages 17-22. Entire text available at: epdf.pub/ berkshire-encyclopedia-of-sustainability-vol-2-the-business-of-sustainability.html.

[25] U.S. Environmental Protection Agency, Emission Factors for Greenhouse Gas Inventories, last modified, April 4, 2014, 5 pages, www.epa.gov/sites/production/files/2015-07/ documents/emission-factors_2014.pdf.

[26] H.P. Van Essen, H.J. Croezen, J.B. Nielsen, Emissions of Pipeline Transport Compared with Those of Competing Modes. Environmental Analysis of Ethylene and Propylene Transport within the EU, CE Solutions for Environment, Economy and Technology (CE-03459832), Oude Delft, The Netherlands, 2003, 118 pages.

[27] T.R. Hawkins, B. Singh, G. Majeau-Bettez, A.H. Strømman, Comparative environmental life cycle assessment of conventional and electrical vehicles, J. Ind. Ecol 17 (1) (2013) 53—64, doi.org/10.1111/j.1530-9290.2012.00532.x.

[28] Engineering-abc.com. Energy consumption during cycling. www.tribology-abc.com/calculators/cycling.htm.

Chapter 6

Buildings

Numbers in this chapter pertain to buildings after their construction. See Section 12.3 for numbers relating to the construction of buildings.

6.1 Sizes

6.1.1 Single-family house

In the United States, the average single-family detached house has 2,856 ft^2 (265 m^2) of floor space [1] while the average household is 2.52 people [2], corresponding to 1,133 ft^2/person (105 m^2/person).

6.1.2 Commercial buildings

Commercial buildings come in all types of sizes for a variety of uses. In the United States, the average building size varies from 4,800 ft^2 (446 m^2) for food services like restaurants to 247,800 ft^2 (23,000 m^2) for health care facilities, with an overall average is 15,700 ft^2 (1,460 m^2) [3]. The following table provides sizes and amount of space per worker, by building type.

Average size of buildings by type of use			
Building type		Mean square feet for building	Square feet per worker
Education	All schools and colleges	31,500	1,124
	Primary and secondary schools	19,400	977
	Colleges and universities	48,400	1,290
Food	Food stores	7,100	1,067
	Dining	4,800	530

Continued

Data, Statistics, and Useful Numbers for Environmental Sustainability.
https://doi.org/10.1016/B978-0-12-822958-3.00002-9

	All	26,500	546
Healthcare	With beds for inpatients	247,800	555
	Outpatients only	12,100	535
Lodging (hotels, motels)		36,900	1,894
Office complex		22,400	465
Retail	Individual stores	12,400	1,352
	Shopping malls	35,900	1,156
Public assembly		15,800	1,716
Religious worship		11,100	2,295
Warehouse and storage		16,400	1,843

Source: [3].

6.2 Materials

From the environmental perspective, the relevant numbers are those pertaining to the thermal insulation and embodied energy in the building materials.

6.2.1 Thermal resistance values

The so-called R-value of a material is a measure of its thermal resistance and therefore of its energy-saving potential. It is the inverse of the conductivity value (so-called U-value), which is expressed in BTUs/(ft^2 × °F × hour) or W/(m^2 × K). The following table provides R- and U-values for common building materials. Additional numbers may be found in the cited sources.

Thermal resistance of building materials			
Building material		**U-value BTUS/ (ft$^2 \cdot$°F·hour)**	**R-value[1]= 1/U-value**
Wood	Softwood	–	1.41/inch
	Hardwood	–	0.90/inch
	Plywood board	–	1.24/inch
	Particle board—low density	–	1.41/inch
	Particle board—medium density	–	1.06/inch

Cement, concrete	Cement mortar	—	0.20/inch
	Solid concrete	—	0.08/inch
	Block with empty core	0.58—1.03	1.71—0.97
Brick		—	0.20/inch
Gypsum wallboard	(3/8-in thick so-called "drywall")	—	0.34
Glass		—	0.51/inch
Windows	Single glazed	1.14	0.88
	Double-glazed	0.52—0.60	1.92—1.67
	Triple-glazed with argon gas	0.17	5.88
Doors	Without glass pane	0.60	1.67
Insulation	Fiberglass blanket	—	3.4/inch
	Cellulose spray	—	3.5/inch
	Urethane foam	—	5.6/inch
	Expanded polyurethane	—	6.2/inch
	Polyiso foam	—	up to 7.62/inch
Roofing	Asphalt shingle	2.3	0.44
	Slate (1/2 inch thick)	20	0.05
Siding	Vinyl siding	—	up to 4.0
	Stucco	—	0.20/inch
	Wood shingles	—	0.87
Air	Still air film	1.5	0.68
	Space between studs	1.05	0.95

[1]R-values are given here in the units used in the United States. To convert into metric units, multiply tabulated values by 0.176.
Sources: [4,5].

6.2.2 Embodied energy

The embodied energy of a material is the energy that was consumed to produce, transport, and deliver the finished material (ex. steel stud) from raw materials (ex. iron ore) before its use in a building.

According to researchers in Australia, the average house contains about 1,000 GJ of energy embodied in its construction materials. This is equivalent to about 15 years of normal operational energy use. For a house that lasts 100

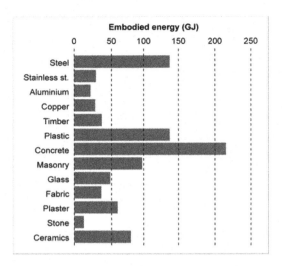

FIGURE 6-1 Breakdown of embodied energy in the various materials used in the construction of a typical house. *Source:* [6].

years, this is over 10% of the energy used in its life [6]. The breakdown is provided in Figure 6-1.

The following table lists the embodied energy of selected building materials on a per-mass basis. The cited source includes additional numbers.

Embodied energy of building materials	
Material	**Embodied energy (MJ/kg)**
Softwood (kiln dried)	3.4
Hardwood (kiln dried)	2.0
Particle board	8.0
Plywood	10.4
Glue-laminated timber	11.0
Laminated veneer lumber	11.0
Plastics (average)	90.0
Polyvinyl chloride (PVC)	80.0
Acrylic paint	61.5
Gypsum plaster	2.9
Cement	5.6
Concrete (poured in situ)	1.9
Clay bricks	2.5

Concrete blocks	1.5
Glass	12.7
Aluminum	170.0
Copper	100.0
Galvanized steel	38.0

Source: [6].

In the United States, 2–7 tons of waste are generated during the construction of a new single-family detached house. This amounts to a rough average of 4 lbs of waste per square foot. Wastes include wood (27%) and other materials such as cardboard, drywall/plaster, insulation, siding, and roofing leftovers, among others. They also include 15–70 lbs of hazardous waste generated construction, including paint, caulk, roofing cement, aerosols, solvents, adhesives, oils, and greases. As much as 95% is clean, unmixed, and recyclable ([7] Section 1.4.12).

6.3 Energy and carbon footprint during use

Comparison of the energy consumption and carbon dioxide emission of the building sector with other sectors of the economy permits the determination of a carbon footprint factor per unit of energy consumed in a building, as tabulated in the following.

Carbon footprint intensity of buildings						
	Residential		Commercial		All buildings	
	Amount	Fraction	Amount	Fraction	Amount	Fraction
Energy consumption (trillion BTUs) [8]	21,483	21.2%	18,439	18.2%	39,922	39.5%
CO_{2eq} emission (million metric tons) [9]	1,017	19.3%	891	16.9%	1,908	36.2%
Emission factor g CO_{2eq}/BTU	0.0473		0.0483		0.0478	

We note that commercial buildings have a slightly higher carbon footprint than residential buildings. The emission factors calculated earlier represent the on-site emissions; when including the primary sources of energy and carbon emissions along the way, the emission factors more than double to 0.1056 g CO_{2eq}/BTU for residential buildings, 0.1187 g CO_{2eq}/BTU for commercial buildings, and 0.1114 g CO_{2eq}/BTU for all buildings combined ([7] Section 1.4.8). This is due in large part to the inefficiencies of generating electricity from primary sources of energy.

6.3.1 Single-family house

The relative energy needs of a single-family detached house in the United States are tabulated in the following.

Energy use in a US house	
Purpose	**Fraction**
Space heating	15%
Air conditioning	17%
Appliances[1]	31%
Water heating	14%
Lighting	10%
Other	13%
Total	100%

[1]Including refrigerator, washer, dryer, and televisions.
Source: [10].

The following table gives the breakdown of energy consumption by fuel type.

Average household energy consumption by fuel type				
(in million BTUs per year; for the United States in 2015)				
Fuel oil	Natural gas	Propane	Electricity	Total
78.2	69.6	35.8	42.7	226.3
35%	31%	16%	19%	100%

Note: Data represent on-site or delivered energy. Consumption of biomass (e.g., wood), coal, solar, and outdoor propane grills are not included.
Source: [11].

Per house, the annual energy consumption is 106,600 BTUs/household or 55,400 BTUs/ft^2 or 39,400 BTUs/person ([7] Section 2.1.11). This translates into 31.2 kWh/household, 175 kWh/m^2, and 11.5 kWh/person annually.

According to estimations by Carbonfund.org [12], the individual carbon footprint attributed to one's housing situation is as follows:

— Apartment (up to 1,000 ft^2): 11,000 lbs (=4,990 kg) of CO_2 per year;

— Small house (up to 1,500 ft^2): 16,500 lbs (=7,480 kg) of CO_2 per year;

— Medium house (up to 2,500 ft^2): 27,500 lbs (=12,470 kg) of CO_2 per year;

— Large house (up to 4,000 ft^2): 44,000 lbs (=20,000 kg) of CO_2 per year.

6.3.2 Commercial buildings

Electricity and natural gas are the two dominant sources of energy in commercial buildings, together accounting for 93% of the energy supply. For 2012, the breakdown was (in trillion BTUs): 4,241 electricity (=1.243 trillion kWh/year), 2,248 natural gas, 341 district heating, and 134 fuel oil, for a total of 6,964 trillion BTUs [13].

Figure 6-2 provides the annual average energy use per square foot in typical commercial buildings in the United States for 2003 and 2012. To translate these numbers into CO_{2eq} emissions, use 0.0483 g of CO_{2eq}

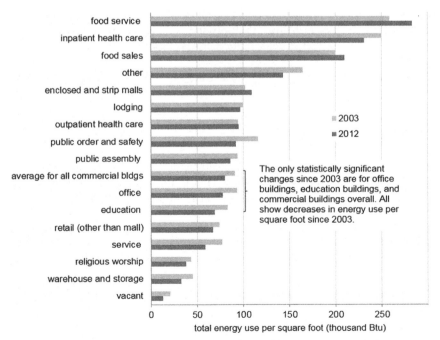

FIGURE 6-2 Energy use in US commercial buildings in 1,000's BTUs per square foot per year. *Source:* [13] Fig. 4.

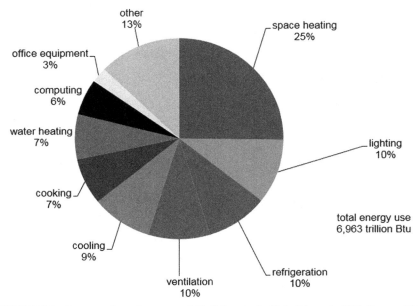

FIGURE 6-3 Energy end uses in commercial buildings in the United States in 2012. *Source:* [13] Fig. 5.

released per BTU consumed in commercial buildings, as calculated earlier. To obtain energy values in MJ/m², multiply the 1,000's BTUs/ft² by 11.36.

The end uses of energy consumption are as depicted in the pie chart shown in Figure 6-3.

A minor footprint of commercial building use is attributable to janitorial services, with an estimation of 49 kg CO_{2eq} per occupant per year ([14] page 252).

6.3.3 Electricity consumption

The electricity consumption of the average house in the United States is 10,972 kWh/year (=1.253 kW) [15]. This is the number to be used when converting amounts of electric power (ex. from photovoltaic panels or wind turbines) into an equivalent number of homes.

For commercial buildings, the annual electricity consumption averages to 14.6 kWh/ft² [16] but varies significantly with the type of activity occurring inside the building, as shown in the following table on a per square foot basis.

Annual electricity consumption in various types of commercial buildings	
Building activity	**Electricity (kWh/ft²)**
Bank	19.3
Education (schools, universities)	11.0

Food sales	48.7
Food service	44.9
Healthcare	25.8
Libraries	15.2
Lodging	15.3
Office	15.9
Post office	9.2
Public order and safety	14.9
Religious worship	5.2
Retail—separate store	15.4
Retail—strip mall	21.7
Retail—enclosed mall	17.4
Vacant	4.5
Vehicle service and repair	8.7
Warehouse	6.6
Warehouse—refrigerated	28.8

Source: [16].

6.4 Water consumption during use

In the United States during 2005, it was estimated that the buildings sector as a whole consumed 39.6 billion gallons per day (bgd) (150×10^9 L/day), consisting of 29.4 bgd (111×10^9 L/day) in residential buildings and 10.2 bgd (38.6×10^9 L/day) in commercial buildings. Over time, the increase in residential water use (1%/year) tracks with population (1.1%/year) whereas consumption in commercial buildings increases almost twice as fast (1.9%/year) ([7] Section 8.1). At an average rate of 2.78 kWh per thousand gallons (7.34×10^{-4} kWh/L) for water treatment, distribution, sewage collection, and wastewater treatment ([7] Section 8.2), the energy footprint of water is 81.7×10^6 kWh/day in residential buildings and 28.3×10^6 kWh/day in commercial buildings.

In the average American home, the indoor water consumption by the household is about 138 gallons per day [17]. This breaks down as tabulated in the following.

Water consumption in a US house	
Indoor use	**Amount (gphd)**
Toilets	33.1 (24%)
Showers	28.1 (20%)

Continued

Faucets	26.2 (19%)
Laundry	22.7 (17%)
Bathing	3.6 (3%)
Dishwashing	1.6 (1%)
Leaks	17.0 (12%)
Other	5.3 (4%)

gphd, gallons per household per day.
Source: [17].

Tabulated in the following is the water consumption in commercial buildings based on the type of building.

Water consumption in commercial buildings		
Building type	**Water use gallons/day**	**Water intensity gallons/(ft^2 × year)**
Hotels and motels	7,113	42.0
Laundry facilities	3,290	—
Car washes	3,031	—
Schools, colleges, and universities	2,117	14.5
Hospitals, medical offices, and other healthcare facilities	1,236	49.7 (inpatient) 15.7 (outpatient)
Office buildings	1,204	14.5
Restaurants	906	—
Food stores	729	—
Auto repair shops	687	—
Religious organizations and other membership-based organizations	629	—
Warehouses and storage	—	3.4
All large buildings	—	20.4

Sources: [7] Section 8.3.2, [18] Fig. 1.

6.5 Demolition

Except for a few antiquities, buildings do not stand forever. The lifetime of commercial buildings in North America is typically in the range of 50–100 years [19]. In midtown Manhattan (New York City), the life expectancy of a large commercial building is 57 years [20]. A building is rarely torn down because of a structural failure, deteriorating physical condition, or fire. In the majority of cases, the cited reason for demolishing a building is the end of usefulness or change in land use.

The amount of energy spent on demolishing a building is negligible compared to the energy required to erect the structure and to operate it during its lifetime. A detailed study of the primary energy spent over the lifetime a house in Michigan [21] and a separate life-cycle study of three different office buildings [22] reach the same conclusion that demolition accounts for no more than a fraction of one percent of the overall energy spent during the life of the building, as tabulated in the following.

Energy and carbon footprint breakdown over the lifetime of a building			
		Average of three office buildings	
	Private home		
Life stage	on energy basis	on energy basis	on CO_2 basis
Manufacturing of materials	6.1%	4.8%	4.2%
Construction		0.2%	0.2%
Operation	93.7%	93.6%	94.7%
Maintenance		1.3%	0.8%
End-of-life	0.2%	0.1%	0.1%

Sources: [21] Fig. 3-7, [22] Fig. 6-2.

In the United States, the amount of debris generated by the demolition of buildings in 2003 was 71.0 million tons comprising 38.0 million tons from residential buildings and 33.0 million tons from commercial buildings ([7] Section 1.4.14). When debris from construction and renovations is added to that of demolition, the number rises to 170 million tons per year, representing approximately 3.2 lbs of solid waste per person per day in the United States ([7] Section 1.4.14).

Sources

[1] U.S. Energy Information Administration (eia), Consumption & Efficiency – 2015 Residential Energy Consumption Survey (RECS) – Square Footage Methodology, October 31, 2017. www.eia.gov/consumption/residential/reports/2015/squarefootage/.

[2] Statista – Society – Demographics – Average number of people per household in the United States from 1960 to 2019. www.statista.com/statistics/183648/average-size-of-households-in-the-us/.

[3] U.S. Energy Information Administration (eia), Consumption & Efficiency — 2012 Commercial Buildings Energy Consumption Survey (CBECS) — Data — Building Characteristics — Table B.1, December 2016. www.eia.gov/consumption/commercial/data/2012/bc/pdf/b1.pdf.

[4] PLTW Engineering, R-Value and Densities Chart. www.windsor-csd.org/Downloads/R-ValueDensitiesChart2.pdf. See also: ColoradoEnergy.org — Insulation Values for Selected Materials. www.coloradoenergy.org/procorner/stuff/r-values.htm.

[5] U.S. Department of Energy, Energy Saver — Energy Efficiency in Log Homes. www.energy.gov/energysaver/types-homes/energy-efficiency-log-homes.

[6] Australian Government, YourHome (Australia's Guide to Environmentally Sustainable Homes) — Materials — Embodied Energy — Authored by Geoff Milne, Updated, 2013. www.yourhome.gov.au/materials/embodied-energy.
[a] E. Adams, J. Connor, J. Ochsendorf, Embodied Energy and Operating Energy for Buildings: Cumulative Energy over Time. Design for Sustainability. Civil and Environmental Engineering, Massachusetts Institute of Technology, Cambridge, MA, 2006.
[b] B. Lawson, Building Materials, Energy and the Environment: Towards Ecologically Sustainable Development, Royal Australian Institute of Architects, Red Hill, 1996. ACT. See also Figures 1 and 2 in.
[c] I.Z. Bribián, A.V. Capilla, A.A. Usón, Life cycle assessment of building materials: comparative analysis of energy and environmental impacts and evaluation of the eco-efficiency improvement potential, Build. Environ. 46 (2011) 1133—1140. https://doi.org/10.1016/j.buildenv.2010.12.002.

[7] U.S. Department of Energy, Energy Efficiency & Renewable Energy — 2011 Buildings Energy Data Book, 286 pages. ieer.org/wp/wp-content/uploads/2012/03/DOE-2011-Buildings-Energy-DataBook-BEDB.pdf.

[8] U.S. Energy Information Administration (eia), Consumption & Efficiency — Current Issues & Trends —Recent Data. www.eia.gov/consumption/.

[9] U.S. Energy Information Administration (eia), Frequently Asked Questions — What are U.S. energy-related carbon dioxide emission by source and sector? www.eia.gov/tools/faqs/faq.php?id=75&t=11.

[10] U.S. Energy Information Administration (eia), Consumption & Efficiency — 2015 Residential Energy Consumption Survey (RECS) — Residential Electricity by End Use. www.eia.gov/consumption/residential/.

[11] U.S. Energy Information Administration (eia), Consumption & Efficiency — 2015 Residential Energy Consumption Survey (RECS) — Table CE2.1 Annual Household Site Fuel Consumption in the U.S. — Totals and Averages, May 2018. www.eia.gov/consumption/residential/data/2015/c&e/pdf/ce2.1.pdf.

[12] Carbonfund.org, Offset Your Life. carbonfund.org/carbon-offsets/.

[13] U.S. Energy Information Administration (eia), Commercial Buildings Energy Consumption Survey (CBES) — 2012 Commercial Building Energy Consumption Survey: Energy Usage Summary. www.eia.gov/consumption/commercial/reports/2012/energyusage/index.php.

[14] A. Martínez-Rocamora, J. Solís-Guzmán, M. Marrero, Carbon footprint of utility consumption and cleaning tasks in buildings. Chap. 9 in: S.S. Muthu (Ed.), Environmental Carbon Footprints — Industrial Case Studies, Butterworth-Heinemann, Elsevier, 2018, pp. 229—258. doi.org/10.1016/B978-0-12-812849-7.00009-X.

[15] U.S. Energy Information Administration (eia), Frequently Asked Questions — How Much Electricity Does an American home Use? www.eia.gov/tools/faqs/faq.php?id=97&t=3.

[16] U.S. Energy Information Administration (eia), Consumption & Efficiency − 2012 Commercial Buildings Energy Consumption Survey (CBECS) − Table PBA4. Electricity Consumption Totals and Conditional Intensities by Building Activity Subcategories, December 2016. www.eia.gov/consumption/commercial/data/2012/c&e/cfm/pba4.php.

[17] W.B. DeOreo, P. Mayer, B. Dziegielewski, J. Kiefer, Residential End Uses of Water, Version 2, Executive Report, Water Research Foundation, 2016, 15 pages.

[18] U.S. Energy Information Administration (eia), Consumption & Efficiency − 2012 Commercial Buildings Energy Consumption Survey (CBECS): Water Consumption in Large Buildings Summary. www.eia.gov/consumption/commercial/reports/2012/water/.

[19] J. O'Connor, Survey on actual service lives for North American buildings, Woodframe Housing Durability and Disaster Conference, Las Vegas, 2004. October 2004, cwc.ca/wp-content/uploads/2013/12/DurabilityService_Life_E.pdf.

[20] New York's Aging Buildings, by Roland Li, Observer, February 1, 2010. observer.com/2010/02/new-yorks-aging-buildings/.

[21] S. Blanchard, P. Reppe, Life Cycle Analysis of a Residential Home in Michigan, Center for Sustainable Systems, University of Michigan, 1998. Report No. 1998-5, September 1998, 71 pages, citeseerx.ist.psu.edu/viewdoc/download?doi=10.1.1.581.5068&rep=rep1&type=pdf.

[22] A.F. Ragheb, Towards Environmental Profiling for Office Buildings Using Life Cycle Assessment (LCA), Ph.D. Thesis, (Architecture), University of Michigan, Ann Arbor MI, 2011, 184 pages, deepblue.lib.umich.edu/bitstream/handle/2027.42/86391/aragheb_1.pdf?sequence=1.

Chapter 7

Electronics and computers

7.1 Integrated circuits (microchips)

Integrated circuits, also called computer chips or microchips, are pieces of silicon wafers with various deposited layers and etchings to form digital circuits in tiny sizes. Although they are the brain of the product, integrated circuits typically make up only about 10% of the total weight of a computer, and their life-cycle carbon emission is at least 1 g of CO_{2eq} per gram of chip, dominated by the carbon emissions from the production of the silicon wafer [1]. The base silicon wafer before etching of the circuits accounts for the majority (slightly upward of 70%) of other forms of impacts according to the Eco-indicators'99 [2]. It takes up to 20 kg of silica and at least 2 MWh (7,200 MJ) of energy to produce 1 kg of sliced and polished silicon wafers [3].

For the 20 mask layers or so, involving up to 200 process steps such as metal deposition, oxidation, etching, and rising (as of 2003), a series of ultra-pure chemicals is necessary, and the whole activity must take place in a Class 1 cleanroom in which constant air filtration consumes 1.7 to 10 MWh/m^2 [3].

By way of illustration, the manufacturing of a 2-g memory microchip takes 32 kg of water, 1.6 kg of fossil fuel, 700 g of elemental gases (mainly N_2), and 72 g of chemicals [4]. The energy breakdown is tabulated in the following.

Energy demand in the production of a 2-gram memory microchip						
Silicon processes	Production of chemicals	Fabrication	Assembly materials	Assembly processes	Lifetime use[1]	Total
5.8 MJ	2.3 MJ	27 MJ	0.17 MJ	5.8 MJ	15 MJ	56 MJ

[1]Assuming 3 hours of use per day for 4 years and accounting for inefficiencies in electricity generation (10.7 MJ per kWh generated).
Source: [4] Fig. 3.

The rule of thumb for energy consumed in the fabrication of microchips suggested by Ref. [4], derived from a multiplicity of studies, is 1.5 kWh of electricity per cm^2 and 1 MJ of fossil fuel energy per cm^2 of the chip surface.

Data, Statistics, and Useful Numbers for Environmental Sustainability.
https://doi.org/10.1016/B978-0-12-822958-3.00010-8

A comprehensive life-cycle assessment of a particular mobile phone mentions that the manufacturing of CMOS *logic* chips causes greater carbon emissions and other impacts than *memory* chips ([5] page 34 [6], Figs. 1 and 2) as well as greater electricity consumption during production (3 vs. 2 kWh per cm^2 of wafer surface according to Section IV-C-2 of [7]).

The environmental impacts of logic chips are as tabulated in the following. Note that numbers are per cm^2 of silicon wafer area and include the upstream impacts of the equipment and general infrastructure used for chip fabrication.

Environmental impacts of logic chips					
Nature of impact	Infrastructure	Silicon wafer	Production of chemicals	Fabrication	Direct emissions and recycling
Energy (MJ/cm^2)	17.9	5.9	2.9	33.6	–
Global warming $(kg\ CO_{2eq}/cm^2)$	1.5	0.5	0.4	0.9	–
Contribution to smog $(g\ NO_x/cm^2)$	7.43	5.25	–	6	0.251
Acidification $(mol\ [H^+]/cm^2)$	0.386	0.303	–	0.356	0.200
Ecotoxicity $(mg\ 2,4\text{-}D/cm^2)$	49.6	26,000	–	30,000	470
Human cancer $(mg\ C_6H_6/cm^2)$	73.6	–	–	–	18.9

Source: [5] Table 3-7.

According to Ref. [8], the Huawei Technologies Company has developed the following formula to estimate the carbon footprint of an integrated circuit based on the mass m_{die} (in mg) of the silicon wafer on which it is etched:

$$\text{Carbon footprint} = m_{die} \times \left(0.0308 \frac{kg\ CO_{2eq}}{mg} + 0.0066 \frac{kWh}{mg} \times 0.6 \frac{kg\ CO_{2eq}}{kWh} \right)$$

In this formula, the first term accounts for manufacturing and the second term for consumption during use, in which the carbon footprint per kWh of electricity is adjustable based on the local sourcing of electricity (stated as 0.6 in the previous formula—see Section 4.3.4 for particular locations).

7.2 Printed circuit boards

Printed circuit boards such as the motherboard of a desktop computer have an estimated energy footprint of 2,830 MJ/kg and a carbon footprint of 162 kg of CO_{2eq}/kg. For laptop computers, the more compact circuit boards yield higher numbers, namely 4,670 MJ/kg and 267 kg of CO_{2eq}/kg ([9] Table 6.3). A separate study quotes a carbon whole-life footprint per unit area of 283 kg CO_{2eq}/m^2 with the manufacturing stage accounting for 57% of this footprint ([10] page 411).

Using the Ecoinvent database widely used for life-cycle assessment analyses of consumer products, one can calculate that the production of a circuit board for a notebook computer is responsible for 55 kg of CO_{2eq} ([8] Section 2.1.1).

For lack of better alternatives, tin-lead solder (60% Sn—40% Pb) continues to be widely used in the assembly of printed circuit boards despite the well-documented fact that lead is a toxic metal. The advantages are good conductivity, high corrosion resistance, and low melting point (188°C, 370°F). It has been estimated that the amount of lead used in soldering is around 50 g/m^2 of printed circuit boards [11] and that a desktop computer with cathode-ray-tube display contains upward of 620 g of lead and 67 g of tin ([12] Table 1). According to a life-cycle assessment study conducted by the US Environmental Protection Agency [13], the environmental impacts of the common lead-tin solder are numerous, with the worse ones tabulated in the following.

Environmental impacts of tin-lead solder				
Nature of Impact		Amount	Unit per liter[1] of solder applied to a circuit board	Mostly during which life-cycle stage(s)
Energy		12,500	MJ	Application
Global warming		817	kg CO_{2eq}	Application
Air particulate matter		45.2	kg	Application
Aquatic toxicity		1,270	kg of aquatic toxic equiv.	End of Life
Occupational health	Cancer	76.2	kg cancer toxic equiv.	Application
	Noncancer	5.60×10^5	kg noncancer toxic equiv.	Manufacturing and End of Life

Continued

Public human health	Cancer	6.96	kg cancer toxic equiv.	Application
	Noncancer	8.80×10^4	kg noncancer toxic equiv.	End of Life

[1]*For volume to mass conversion, the density of lead-tin solder is 8.5 kg/L at room temperature.*
Source: [13] Table 4.1.

7.3 Computers

Life-cycle analyses of laptop and desktop computers have produced inconsistencies [8,14], making it difficult to quote reliable numbers. In particular, some studies (ex. [15,16]) have concluded that the manufacturing phase is far more impacting than the use phase while others have reported the opposite conclusion (ex. [8,17−19]) or determined that both manufacturing and use phases have similar impacts (ex. [20,21]). The factsheet published by the Center for Sustainable Systems at the University of Michigan pegs the energy spent during a 3-year use at 34% of the total in the whole life cycle [22]. Some of the disagreement is due to the consideration of different computers over time and to different assumptions made about use as well as the levels of upstream impact of some materials. Therefore, numbers presented in the following should be adopted according to the level of trust placed in the source.

7.3.1 Total life cycle

The following table gives the life-cycle impacts of computers with either cathode ray tube (CRT) or liquid crystal display (LCD).

Life-cycle impacts of a computer, for two different displays					
	CO_2 emission (kg/functional unit)		Water consumption (L/functional unit)		Solid waste (kg/functional unit)	
Life cycle stage	CRT	LCD	CRT	LCD	CRT	LCD
Upstream[1]	29.2	107	554	263	9.55	13.1
Manufacturing	179	62.2	11,400	2,150	81.2	12.6
Use	445	166	1,140	425	83.3	31.1
End-of-life	2.59	1.39	−27.3	−18	−1.66	−4.42
Total	655	336	13,100	2,820	172	52.3

[1]*Counting material extraction and processing for key materials.*
Source: [23] Tables 2-25, 2-29, 2-45 & 2-49.

The following table provides the consumption of fossil fuel, electricity, and their total during the production and use of computers.

Energy use during the life-cycle of a computer			
	Direct fossil fuel (MJ)	Electricity (kWh)	Total energy (MJ)
Production			
Semiconductors	298	170	909
Printed circuit boards	26.7	7.71	54.5
CRT manufacture and assembly	210	12.5	255
Bulk materials—control unit	N/A	N/A	770
Bulk materials—CRT	N/A	N/A	800
Silicon wafers	N/A	38.1	137
Inputs and outputs			
Chemicals	381	18.5	448
Manufacturing equipment	392	29.4	498
Passive components	109	10.3	146
Disk drives and other parts	365	23	446
Transport	338	3.5	351
Packaging and documentation	120	4.8	137
Other	973	61	1,192
Total Production	3,300	430	6,400
Use[1]	0	420	1,500
Total Production and Use	3,300	850	7,900

[1]Assuming a home user, for 3 years.
Source: [15].

7.3.2 Manufacturing

Personal computers, both desktops and laptops, are produced from a wide variety of materials, with a few contributing most to the total weight of the product, as tabulated in the following.

Materials in a personal computer	
Material	**Percentage of total weight**
Steel	29%
Glass, if any	22%

Continued

Plastics	19%
Aluminum	10%
Lead	8%
Copper	7%
Zinc	3%
Other	3%
Source: [22].	

The following table lists the relative abundance of material groups on a mass basis, broken down by main computer processing unit, display unit, excluding the battery if any.

Material composition of the major components of a personal computer

	Ferrous metals	Aluminum	Copper	Other metals	Plastics	Glass	Liquid crystal[1]	Printed circuit board
Computer processing unit	59%	11%	4%	<1%	12%	—	—	14%
Cathode-ray tube (CRT) display	5%	1%	3%	<1%	19%	61%	—	11%
Flat-panel display	37%	7%	1%	<1%	22%	—	26%	6%

[1]Consisting of various chemical elements, some metallic and others non-metallic.
Source: [24].

The metals most used in the manufacturing of computers contribute variously to climate change, with gold being a much more impacting constituent than other metals, as indicated by the environmental scores tabulated in the following. Very little gold, however, is used in a desktop computer and its display screen, 0.39−0.67 g according to Ref. [12].

Global warming potential of metals in a personal computer

Metal	Gold	Silver	Tin-lead solder	Tin	Aluminum	Nickel	Copper
GWP100 per kg of metal[1]	13,600	150	96	21	19	12	4
Eco-Indicator'99 score[2]	>25,000	95	—	20	1	9	11

[1]GWP100 is an estimation of the global warming potential over 100 years from the production of 1 kg of the material compared to 1 kg of carbon dioxide.
[2]The Eco-Indicator'99 score is a number that expresses the total environmental load of a product, as used in standard life-cycle assessment tools.
Sources: [13] for lead-tin solder, [1] for other metals.

The total energy and fossil fuels used in producing a desktop computer in the 1990s with 17-inch CRT monitor were estimated at 6,400 MJ and 260 kg, respectively. This indicates that computer manufacturing is energy-intensive: the ratio of fossil fuel use to product weight is 11, an order of magnitude larger than the factor of 1−2 for many other manufactured goods. This high energy intensity of manufacturing, combined with rapid turnover in computers, results in an annual life cycle energy burden that is surprisingly high: about 2,600 MJ per year, 1.3 times that of a refrigerator. In contrast to various home appliances, the life-cycle energy of a computer is primarily due during manufacturing (81%) rather than during use (19%) [15].

A review of 20 life-cycle assessment studies concludes that the embodied carbon footprint of a desktop computer ranges from 200 to 800 kg CO_{2eq} and from 100 to 400 kg CO_{2eq} for a laptop [25].

The following table lists the estimates of the amounts of water, fuel, and chemicals used in the production of a desktop computer. Omitted numbers were deemed negligible in comparison to others.

Water, fuels, and chemicals used in the production of a desktop computer			
Step in manufacturing	Water (kg)	Fossil fuels (kg)	Chemicals (kg)
Silicon wafers	−	17	−
Semiconductors	310	94	7.1
Printed circuit boards	780	14	14
Cathode-ray tube display	450	9.5	0.49
Bulk materials for control unit	−	21	−
Bulk materials for display	−	22	−
Electronic materials and chemicals (excl. wafer chips)	−	64	−
Manufacturing of larger parts	−	−	−
Assembly	−	−	−
Total	1,540	242	21.6

Source: [26] Table 9.

Figure 7-1 shows the relative environmental impacts of a personal computer according to their type: ADP = abiotic depletion, Acid = acidification of water, rain and snow, GWP = global warming potential, ODP = ozone depleting potential in stratosphere, Eut = water eutrophication, POCP = photochemical ozone creation potential, HT = human toxicity, and ET = ecotoxicity.

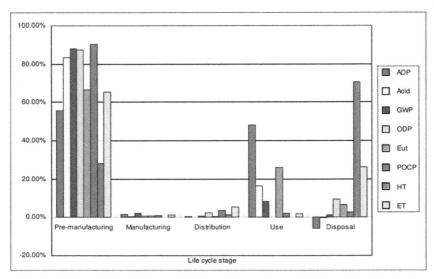

FIGURE 7-1 Relative environmental impacts of a personal computer. See text for the meaning of acronyms. *Source:* [16] Fig. 2.

Thanks to their more compact format, laptop computers (also called notebooks) tend to have slightly lower impacts than desktop computers (Abstract of [27]). According to a study [28] that compared 11 different laptop computers dated from 1999 to 2008 with a focus on their main circuit boards ("motherboards"), there is an average of 57 microchips per board, and the average size of a microchip is 20 mm^2 (including its resin package).

The following table summarizes both the energy demand and carbon emissions of a typical laptop circa 2010.

Energy demand and carbon footprint of a laptop					
		Energy (in MJ)		CO$_2$ emission (in kg)	
Life-cycle stage		Low	High	Low	High
Upstream	Chemicals	331		32	
	Equipment	292		18	
	Indirect[1]	1,070		83	

Continued

Fabrication	Silicon wafer	60		5	
	Integrated circuits	247	405	21	33
	Circuit board	30	43	2	3
	LCD display	264	932	13	6
	Additional materials	280	665	16	41
	Assembly	435	541	35	47
Total for production		3,009	4,339	227	270
Use		1,781		159	
Total for production + use		4,790	6,121	386	429

[1]Upstream of upstream as estimated from an economic input–output life-cycle assessment (EIO-LCA) calculation.
Source: [27].

The amounts of energy, water, and chemicals used in the manufacture of LCDs are tabulated in the following.

Water, electricity and materials used in the production of a liquid crystal display	
Material or input	**Amount used per monitor**
Water	1,290 L
Photolithographic[1] and other chemicals	3.7 kg
Elemental gases (nitrogen, oxygen, argon)	5.9 kg
Direct fossil fuels (mostly natural gas)	198 kg
Embodied fossil fuels	226 kg
Electricity	87 kWh

[1]Photolithographic etchants include chlorine (Cl_2), hydrochloric acid (HCl), boron trichloride (BCl_3), carbon tetrafluoride (CF_4), carbon trifluoride (CHF_3), and sulfur hexafluoride (SF_6).
Sources: [23] Section 2.7.1.2, as summarized in Ref. [26] Table 4.

The global warming potential incurred during the manufacture of a laptop tends to be an increasing function of its weight, with the production of a 1.5 kg laptop being responsible for about 150 kg of CO_{2eq} and a 3.0 kg laptop responsible for about 450 kg of CO_{2eq}, although there are major exceptions ([29] Fig. 1).

According to a 2011 comprehensive report ([30] particularly its Section 6.2), mining activities and their fossil fuel consumption are the chief contributors to the environmental impacts of a laptop. The reader is referred to this report for details.

7.3.3 Use

In Europe, according to Ref. [8], a laptop computer used for 5 years in an office consumes a total of 580 kWh, which amounts to 1,600 kWh (=5,760 MJ) of primary energy, compared to 66 kWh for the electricity consumption during its manufacturing. Citing a different study using the Ecoinvent database [8], mentions a "typical laptop" of 3.2 kg used during 4 years uses 190 kWh, and its main circuit board alone is responsible for 55 kg of CO_{2eq} [8].

In South Korea, a personal computer used at home is operational for 10.2 hours of the week and idle for another 3.2 hours; in its lifetime of 4 years, it consumes a total of 197 kWh. At the office, the numbers rise to 12.9 hours in operation and 8.95 hours in idle mode per week, for a total consumption of 305 kWh during a 4-year lifespan [16].

In the United States, according to Ref. [22], the average desktop computer consumes 48 W in use or idle and 2.3 W in sleep mode, for an annual total of 285 kWh, whereas laptops consume only about 15 W in use or idle and 1.2 W in sleep mode, for an annual total of 89 kWh. The most consuming component is the lighted screen display; a 17-inch liquid-crystal display requires 13 W while on, 0.4 W in standby mode, and still 0.3 W when turned off [22].

The carbon emission per time of use has been estimated at 6–6.9 g CO_{2eq} per hour for the Apple Macs and to 1.5 g of CO_{2eq} per hour for the Apple iPad tablet ([31] Table 3).

7.3.4 End of life

When a piece of electronic equipment is no longer wanted or needed by its user; it can be reconditioned and reused, or it can be recycled, or it can be simply discarded.

7.3.4.1 Reuse

Reconditioning followed by reuse is the far preferable option. It not only adds longevity thereby reducing the need to manufacture new units, but the refurbishing company also steers nonreusable parts into a recycling stream [32]. The environmental impacts of reconditioning activities are negligible [32], and the benefits of reuse depend only on the capture rate, say R, and the duration of the life extension, say E as a fraction of the length of the first use duration. If the impact of one material or one step in manufacturing was $I_{initial}$, then the impact of that same material or step after reuse is

$$I_{with\ reuse} = \left(1 - R + \frac{R}{1+E}\right)I_{initial}.$$

The benefit reduction is uniformly spread across all impact categories.

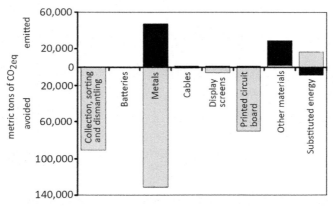

FIGURE 7-2 Climate change impact of recycling electronics in Switzerland. Dark bars refer to impacts occurring at the recycling facility whereas light-shaded bars indicate indirect impacts due to materials avoided or needed upstream. *Source:* [34] Fig. 6, redrawn for added clarity.

If the energy spent on reconditioning the equipment (desktop computer or mobile phone) is substantial (on the order of 30% of that of the original manufacturing) and if the extended lifetime is too short (on the order of 25% or less of the time of first use), the refurbished product is less energy efficient than the original product, and recycling would have been preferable [33]. Likewise, a new laptop is more energy-efficient than a refurbished desktop, and a new LCD is preferable to a refurbished CRT display [18].

7.3.4.2 Recycling

Considerations of waste electrical and electronic equipment show that the benefits of recycling largely exceed its impacts (Figure 7-2). The chief benefits are those of the salvaged materials, their embodied energy, and the avoided burden of primary production [34].

A detailed study of an electronics recycling facility in Belgium led to the following percentages of materials on the input side and their recycling fraction. Except for precious metals and miscellaneous metals, desktop computers are more intensely recycled than laptops.

Material inputs for two types of personal computers				
	Desktop computers		Laptop computers	
Material	Fraction of total input	Fraction recycled	Fraction of total input	Fraction recycled
Ferrous metals	37.2%	89%	14.2%	86%
Aluminum	4.61%	83%	8.44%	75%

Continued

Copper	4.32%	78%	6.85%	85%
Precious metals	0.0113%	49%	0.0290	63%
Other metals	0.639%	29%	10.9%	90%
Plastics	18.8%	43%	40.6%	13%
Other organics	0.0914	0%	0.074	0%
Minerals	30.0%	0%	12.6%	0%
Other materials	4.36%	0%	6.32%	0%
	Total 100%	Weighted average 49%	Total 100%	Weighted average 39%

Source: [35] Table 1.

7.3.4.3 Disposal

One form of disposal of electronic products is their incineration, which is noticeably more impacting than recycling and also worse than landfilling in terms of carbon emissions. If the electronic waste is incinerated in a specially designed facility for this purpose, the global warming potential impact is still 3.7 times greater than for recycling and 22% greater than for landfilling ([36] Fig. 4, GWP panel).

Leaching of toxic materials from computers in well-managed landfills is very small, and landfilling in a highly developed country is far preferable than informal recycling in a developing country [12].

Additional numbers on waste disposal of electrical and electronic equipment can be found in Chapter 11.

7.4 Personal electronics

7.4.1 Smartphones

A life-cycle assessment of a Sony Mobile Z5 (circa 2015) revealed the following carbon footprints, with the production phase being by far the greatest contributor (79%).

Carbon footprint of a mobile phone	
Life cycle stage	Carbon footprint (kg CO_{2eq})
Raw materials	1.9

Continued

Production (including intermediate transport)	Integrated circuits (microchips)	32.8	44.9
	Printed circuit board	2.1	
	Display	3.5	
	Battery	1.4	
	Other parts	2.7	
	Final assembly and vendor activities	2.4	
Final transport		3.0	
Use[1]		7.2	
End of life[2]		−0.3	
Total		56.7	

[1]Assuming 3 years of average use at 4 kWh/year from an electricity mix with 0.6 kg CO_{2eq}/kWh.
[2]Assuming 83% recycling and 17% landfilling.
Source: [7].

A detailed life-cycle assessment conducted in Germany on the Fairphone 2 (circa 2016) estimated the various types of impacts as tabulated below, with the following assumptions for the use phase: 7 hours of use per day for 3 years, charger efficiency of 69%, and no-load charger power loss of 30 mW.

Life-cycle assessment of a mobile phone					
Impact category	Global warming potential	Abiotic resource depletion		Human toxicity	Ecotoxicity
		Materials	Fossil fuels		
Units	kg CO_{2eq}	g Sb_{eq}	MJ	kg DCB_{eq}	kg DCB_{eq}
Production	35.98	1.48	148.03	8.35	0.107
Transport	3.00	0.605×10^{-3}	43.01	2.32	0.00318
Use	5.98	2.66×10^{-3}	63.46	0.24	0.0050
End of life	−1.11	−0.871	−12.85	−0.81	−0.00687
Total	43.85	0.612	241.65	10.11	0.108

Source: [5] Table 4-2.

Several other life-cycle assessments of mobile phones have been conducted that reveal a certain pattern although various studies do not converge on the same numbers, as tabulated in the following. Clearly, the production and transport phase causes the most emissions, and the lifetime carbon emissions weigh significantly more than the product itself.

Comparative impacts of mobile phones					
Regional relevance		Finland	Global	Sweden/ China	Average
Weight of phone		80 g	250 g	80 g	—
CO_{2eq} emission over lifetime		14.4 kg	30 kg	20 kg	21.5 kg
CO_{2eq} emission per gram of product		180 g/g	120 g/g	250 g/g	180 g/g
Contributions to CO_{2eq} emissions	Production and transport	71%	93%	80%	81.3%
	Use	29%	13%	20%	20.7%
	End of life	0%	−7%	0%	−2.3%
Source: [1].					

A mobile phone consumes 0.3 W in average ([25] Table 1), less during idling and more during use. The carbon emission per time of use has been estimated at 1.1 g CO_{2eq} per hour for the Apple iPhone 5 ([31] Table 3).

The overall recycling rate of smartphones in Europe is estimated at 83% with the remaining 17% going to the landfill [7]. Repurposing is also an option: After replacement of the battery, an obsolete smartphone may be converted into a handset to run a parking meter app [37].

7.4.2 Wearables

Wearable electronics are small electronic devices worn by a person to provide intelligent assistance. They can be implanted inside the body (ex. pacemakers and neuroprosthetics), but most are worn externally in the form of a wristwatch or wristband and are sometimes integrated into eyeglasses, clothing, or fashion accessories. Common types are the smartwatch and activity tracker. At the time of this writing, still little is known about the environmental impacts of wearable electronics aside from their electricity consumption, which may be nil if the wearable is equipped with its own mini energy harvesting system [38].

The preliminary observations are rather speculative. When combined with clothing or accessories like handbags, electronic wearables may be as short-lived as fashion, or on the other hand, they may have the effect of prolonging the use of the clothing or handbag. Their small size could also cause them to be more readily thrown away with common household waste [39].

Possibly, the environmental effects of wearables may be more indirect than direct, in the sense that the environmental impact of the modified behavior they induce on the part of the wearer may be greater than that of the device itself. For example, a health tracker could promote better health by exercise and thus avoid the environmental impacts of acute health care. Electronic wearables may harvest low-power from body movement offsetting the need for batteries in other electronics. They could also feed data such as air humidity and pollutant levels into a distributed environmental monitoring network, with downstream benefits.

7.5 Other electronic equipment

7.5.1 Computer accessories

The following table provides estimates of the energy consumed and carbon emission produced by common computer accessories.

Energy demand and carbon footprint of computer accessories		
Accessory	Energy consumed (in MJ)	CO_{2eq} emission (in kg)
Cable, per meter	5–10	0.2–0.5
Hard disk drive	65–216	4–12
Keyboard	468	27
Lithium-ion battery (per kg)	900–935	74–100
Mouse (optical, with cable)	93	5
Power supply	574	30
Source: [9] Table 6.3, pages 129–130.		

7.5.2 Printers and photocopiers

There are three types of printers, the low-cost ink-jet printers usually used in the home, the laser printers typically found in offices, and the more recent solid-ink printer. Each has its distinctive footprint. The following numbers do not include the impact of the paper consumed by these printers but include the ink and its cartridges.

A typical ink-jet printer consumes 20 W when printing and 4 W when idling. It is capable of printing 16 pages per minute. If the printer is used 1 minute per day (for the printing of 16 pages) and is idling 3 hours per day over a 3-year lifetime, the electricity consumption amounts to 18.25 hours of printing at 20 W and 3,285 hours of idling at 4 W, for a total of 13.5 kWh (48.6 MJ) [40]. For this, it consumes 50 ink cartridges. Estimates of energy consumption for manufacturing, transport (from Asia to Europe), ink life cycle, and disposal are 14.9, 0.812, 4.37, and 0.05 MJ, respectively. In terms of climate change impact over a 3-year lifespan, manufacturing, electricity consumption, and paper usage are comparable [40].

The production of a laser printer requires 907 MJ of energy and emits 68 kg of CO_{2eq}, and one toner module is responsible for 215–220 MJ of energy and 10–11 kg of CO_{2eq} ([9] Table 6.3).

Lexmark International performed a life-cycle assessment of its Energy Star® color laser printer, which weighs 62.4 kg (breakdown tabulated in the following) and can print 60 pages per minute. The energy spent in manufacturing is 540 MJ (8.65 MJ/kg in average) [41]. Based on a lifetime of 5 years with an output of 1,790 pages per day and based on 66.7% capture of old printers by the company for recycling, the life-cycle impacts of this printer were estimated as tabulated in the following. Use is the life stage that dominates every type of impact (see details of report [41]).

Life-cycle assessment of a color laser printer						
Material		Mass (kg)	Nature of impact	Unit	per 1,000 pages	for lifetime
Plastics	recyclable	16.7	Global warming	kg of CO_2 equiv.	1.94	4,530
	nonrecyclable	5.34				
Metals	Ferrous	31.5	Ozone depletion	kg of CFC-11 equiv.	5.47×10^{-9}	1.27×10^{-5}
	Aluminum	0.914				
	Copper	0.111	Acidification potential	kg of SO_2 equiv.	0.00719	16.7
Glass		0.272				
Electronics		7.16	Fossil fuel depletion	kg of oil equiv.	0.763	1,780
Other materials		0.388				
Total		62.39	Mineral resource depletion	kg of Fe equiv.	0.212	493

Characteristics of the Lexmark CS820DE laser printer.
Source: [41].

Solid ink printers have a significantly lower footprint than laser printers, in part because solid ink does not require a cartridge. Their cumulative energy demand is about a 34% lower, their global warming potential 30% lower, and their post-consumer waste about 90% lower [42].

In a first approximation, office photocopiers may be assimilated to large laser printers, like the Lexmark printer described earlier, with impacts prorated to their total weight. The Center for Sustainable Systems at the University of Michigan estimates that a photocopier consumes 220 W while in use and 30 W when idling [22].

A study of end-of-life Xerox^{TM} photocopiers calculated the reduction in impacts following disassembly and remanufacturing. As tabulated in the following, savings can be significant especially if the original product was designed with modularity to facilitate disassembly and remanufacturing [43]. The listed number should be indicative of gains to be made by disassembly and remanufacturing of other large electronic equipment with mechanical parts such as 3D printers.

Possible reduction factors for large electronic equipment with mechanical parts		
Type of savings	Reduction factor for nonmodular design	Reduction factor for modular design
Materials (kg)	1.3	1.9
Energy (MJ)	1.4	3.1
Water (L)	1.2	1.6
Landfilled waste (kg)	1.5	1.9
CO_{2eq} (kg)	1.3	2.9
Source: [43] Table 1.		

7.5.3 Display screens and televisions

Flat-panel LCDs have between 2,950 and 3,750 MJ/m^2 of embodied energy and cause emissions between 295 and 375 kg of CO_{2eq} per m^2 during production ([9] Table 6.3).

The carbon emission per time of use of the Apple TV consuming 3 W when streaming movies has been estimated at 1.4 g CO_{2eq} per hour ([31] Table 3). Apple Inc. estimates that its Apple TV 4K has a total carbon footprint of 58 kg CO_{2eq}, made of 60% (34.8 kg) in production, 9% (5.2 kg) in transport, 29% (16.8 kg) in use, and 2% (1.2 kg) in recycling [44].

Although much less bulky than their preceding CRTs, LCDs do not have lower carbon footprints, 1,100 kg CO_{2eq} per LCD television versus 1,050 kg CO_{2eq} per CRT television ([8] Table 5).

According to Ref. [8], a typical television is used between 8 and 10 years during which it consumes around 213−290 kWh/year.

A thorough life-cycle assessment of a 42-inch (0.486 m^2) 30.2 kg plasma television yielded, among other numbers, the ones tabulated in the following. Clearly, the use phase dominates the impacts.

Impacts of a plasma screen television					End of life		
Impact	Unit	Production	Distribution[1]	Use[2]	Effort	Benefit[3]	Total
Primary energy[4]	MJ	12,240	305	46,980	276	−2,584	57,220
CO_2 emissions	kg	637	17.8	2,040	33.1	−125	2,603
Methane emissions	kg	1.12	0.0238	3.54	0.0303	−0.221	4.49

[1]Based on 2,500 km by truck.
[2]Based on 303 W consumption for 4 hours of use per day during 8 years, with 1.6 W consumption in stand-by mode, with adjustments for holidays.
[3]Assuming recycling of metals and energy recovery from incineration.
[4]In terms of coal, crude oil, natural gas, and clean electricity.
Source: [45].

7.5.4 Radios, digital cameras, headsets, and other consumer electronics

The growing category of consumer electronics consists of everyday objects in one's home, including rather mundane appliances in the kitchen in which are now embodied electronic chips and printed circuit boards such as microwave ovens, toasters, and coffee makers. Small electronics are also found in the bathroom (hairdryers and electric toothbrushes with induction chargers), in the toolbox (cordless drills, voltmeters), in the nursery (some animated toys), and in the bedroom (nightstand radio alarms). Relatively new on the scene are noise-canceling headsets and smart thermostats. As a rule of thumb, small electronic devices have between 2 and 4 MJ of embodied energy per gram of product and cause emissions between 200 and 400 g of CO_{2eq} per gram of product ([9] Table 6.3).

As such, the environmental impacts of small electronics do not differ much from those of a tablet minus the LCD screen and, for some, more metallic parts or electricity-consuming heat-generating parts. For the larger ones, the life-cycle stage with the highest impacts is the use phase caused by the electricity consumption of the object whereas for the smaller ones the procurement of materials and their processing are the most impacting elements (see Figure 7-3). For the smallest of objects, packaging may not be negligible.

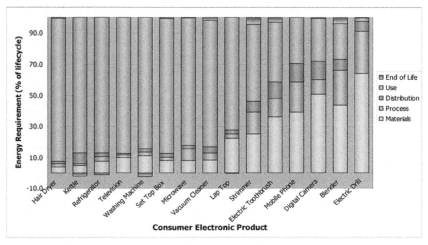

FIGURE 7-3 Partitioning of the energy requirements for a group of consumer electronics. *Source:* [46] Fig. 10.

For a product for which the use phase is dominant, the behavior of the consumer is key, and assumptions made about it, such as frequency of use and lifespan, become crucial.

The carbon footprint of an induction cooker is estimated to be 33.6 kg CO_{2eq} [47].

7.5.5 Musical instruments and music players

At the time of this writing, no environmental data appear to be available on the production and disposal of electronic musical instruments such as electronic keyboards. Energy during use, however, can be estimated rather simply from the wattage of the equipment times the number of hours it is used per week or per year. For example, a CD player is estimated to use 34.7 kWh per year [48].

The carbon footprint then follows by the multiplication of the energy consumed by the carbon footprint of the local electricity mix (0.45 kg CO_{2eq}/kWh in the United States).

7.5.6 3D printers

3D printers are pieces of equipment used for so-called additive layer manufacturing. While they are currently used mostly for prototyping, the expectation is that they will provide an alternative method of manufacturing on a large scale. Their advantages are material efficiency, part flexibility (one complicated part instead of simpler parts to be assembled), and production flexibility [49], all of which translate into decreased environmental impacts. The Center for Sustainable Systems at the University of Michigan [22] predicts that 3D printers have the potential for a major reduction in total primary energy use and CO_2 emissions over conventional forms of manufacturing, estimated at 2.5−9.3 EJ (about 1% of world energy consumption in 2018) and 131−526 Mt of CO_2 (about 1% of global emissions in 2019), respectively.

The major environmental impact of a 3D printer does not lie in its production but in its use due to the heat necessary to melt the material used to form the object layer by layer. A practical, though not perfect, metric is the amount of energy spent per kg of object printed. An estimate based on the performance of four different fused deposition printers is 162 kWh/kg, which is higher than for alternative methods of rapid prototyping (31.5 kWh/kg for stereolithography and 35.0 kWh/kg for laser sintering) [50].

Faludi and collaborators [49] compared the life-cycle impacts of two 3D printers to those of a conventional Computer Numerical Control (CNC) milling machine, considering two extreme scenarios: minimal utilization with the fabrication of two objects per week and idling the rest of the time, and maximal utilization with non-stop fabrication. The following table provides the relative impacts, showing clearly that electricity consumption is the dominant source of impact of all machines, and more so for 3D printers than for the conventional milling machine.

Relative energy demand for selected 3D printers						
	Dimension FDM 3D printer		**Object Connex 350 3D printer**		**Conventional CNC milling machine**	
Lifetime	5 years		5 years		10 years	
Life-cycle stage	Minimum utilization	Maximum utilization	Minimum utilization	Maximum utilization	Minimum utilization	Maximum utilization
Manufacturing	5.6%	4.7%	24.6%	7.1%	12.2%	8.5%
Transport	0.76%	0.94%	6.59%	2.09%	0.86%	0.57%
Materials	0.072%	0.00%	0.29%	0.00%	0.49%	0.57%
Electricity	92.4%	80.2%	66.3%	68.6%	82.2%	35.1%
Waste produced	1.08%	11.32%	1.49%	15.06%	0.95%	17.56%
Lubrication oil	0.11%	2.83%	0.74%	7.11%	3.23%	36.54%
Disposal	−	−	−	−	0.06%	1.13%
Total	100%	100%	100%	100%	100%	100%

Source: [49] Fig. A4, with score points converted into percentages.

In a subsequent study, Faludi and other collaborators [51] performed cradle-to-grave life-cycle assessments of six 3D printers and concluded that the choice of material being used in the printing is paramount. The lower its melting point, the better; best are room-temperature powders that can be bonded with low-toxicity adhesives.

7.5.7 Toys

Toys vary widely in size and variety of electronic features. A study of a 1,047-g plush teddy bear with 167 g of electronics and 72 g of batteries concluded that its carbon footprint, including packaging of 313 g is 19 kg CO_{2eq} (=18 times the mass of the toy), but that this number could be reduced to 1.3 kg CO_{2eq} (about the mass of the toy) by decreasing the electricity demand, switching to rechargeable batteries, minimizing the packaging, and using recycled fibers for the stuffing [52].

Sources

[1]   O. Andersen, J. Hille, G. Gilpin, A.S.G. Andrae, Life Cycle Assessment of Electronics. Sust. Tech. 2014 Conference, Portland OR, 2013, pp. 24–26. July 2014, 8 pages.

[2]   A.S.G. Andrae, O. Andersen, Life cycle assessments of integrated circuit packaging technologies, Int. J. Life Cycle Assess. 16 (3) (2011) 258–267, doi.org/10.1007/s11367-011-0260-3.

[3]   A. Plepys, The environmental impacts of electronics. Going beyond the walls of semiconductor fabs, in: ISEE '04. Proc. Intl. Symposium on Electronics and the Environment, 2004, pp. 159–165.

[4]   E.D. Williams, R.U. Ayres, M. Heller, The 1.7 kilogram microchip: energy and material use in the production of semiconductor devices, Environ. Sci. Technol. 36 (24) (2002) 5504–5510, doi.org/10.1021/es0256430.

[5]   M. Proske, C. Clemm, N. Richter, Life Cycle Assessment of the Fairphone 2, Final Report, Fraunhofer IZM, Berlin, Germany, 2016, 73 pages, www.fairphone.com/wp-content/uploads/2016/11/Fairphone_2_LCA_Final_20161122.pdf.

[6]   M. Schmidt, H. Hottenroth, M. Schottler, G. Fetzer, B. Schlüter, Life cycle assessment of silicon wafer processing for microelectronic chips and solar cells, Int. J. Life Cycle Assess. 17 (2) (2012) 126–144, doi.org/10.1007/s11367-011-0351-1.

[7]   M. Ercan, J. Malmodin, P. Bergmark, E. Kimfalk, E. Nilsson, Life cycle assessment of a smartphone, in: Advances in Computer Science Research, Proceedings of 4th Int. Conf. on ICT for Sustainability (ICT4S 2016), 2016, pp. 124–133, doi.org/10.2991/ict4s-16.2016.15.

[8]   A.S.G. Andrae, O. Andersen, Life cycle assessment of consumer electronics – are they consistent? Int. J. Life Cycle Assess. 15 (8) (2010) 827–836, doi.org/10.1007/s11367-010-0206-1.

[9]   M.F. Ashby, Materials and the Environment – Eco-Informed Material Choice, second ed., Butterworth-Heinemann, 2013, 616 pages.

[10] W.K.C. Yung, S.S. Muthu, K. Subramanian, Carbon footprint analysis of printed circuit board. Chap. 13, in: S.S. Muthu (Ed.), Environmental Carbon Footprints – Industrial Case Studies, Butterworth-Heinemann, Elsevier, 2018, pp. 365–431, doi.org/10.1016/B978-0-12-812849-7.00013-1.

[11] S. Herat, Green electronics through legislation and lead free soldering, Clean Soil Air Water 36 (2) (2008) 145–151, doi.org/10.1002/clen.200700164. core.ac.uk/download/pdf/143865862.pdf.

[12] E. Williams, R. Kahhat, B. Allenby, E. Kavazanjian, J. Kim, M. Xu, Environmental, social, and economic implications of global reuse and recycling of personal computers, Environ. Sci. Technol. 42 (17) (2008) 6446–6454, doi.org/10.1021/es702255z.

[13] U.S. Environmental Protection Agency, Solders in Electronics: A Life-Cycle Assessment Summary, EPA-744-S-05-001, August 2005, 51 pages, www.epa.gov/sites/production/files/2013-12/documents/lead_free_solder_lca_summary.pdf.

[14] P. Teehan, M. Kandlikar, Sources of variation in life cycle assessments of desktop computers, J. Ind. Ecol. 16 (s1) (2012) S182–S194, doi.org/10.1111/j.1530-9290.2011.00431.x.

[15] E. Williams, Energy intensity of computer manufacturing: hybrid assessment combining process and economic input-output methods, Environ. Sci. Technol. 38 (2) (2004) 6166–6174, doi.org/10.1021/es035152j.

[16] B.-C. Choi, H.-S. Lin, S.-Y. Lee, T. Hur, Life cycle assessment of a personal computer and its effective recycling rate, Int. J. Life Cycle Assess. 11 (2) (2006) 122–128, doi.org/10.1065/lca2004.12.196.

[17] C. Hermann, Environmental footprint of ICT equipment in manufacture, use and end of life, in: ECOC '08, 34th European Conference and Exhibition on Optical Communication 2008, 2008.

[18] S. Sahni, A. Boustani, T.G. Gutowski, S.C. Graves, Reusing personal computer devices – good or bad for the environment?, in: Proc. 2010 IEEE International Symposium on Sustainable Systems and Technology, Arlington VA, 2010, pp. 1–6.

[19] M. Stutz, Carbon Footprint of a Typical Rack Server from Dell, Company Communication, 2011. Dell Corporation, 4 pages, i.dell.com/sites/content/corporate/corp-comm/en/documents/dell-server-carbon-footprint-whitepaper.pdf.

[20] a M. Eugster, R. Hischier, H. Duan, Key Environmental Impacts of the Chinese EEE-Industry: A Life Cycle Assessment Study, Joint Report from Technology & Society Laboratory, EMPA - Swiss Federal Laboratories for Material Science and Technology (Switzerland) and Tsinghua University, China, 2007;
b See also, M. Eugster, R. Hischier, Key Environmental Impacts of the Chinese EEE-Industry – A Life Cycle Assessment Study, Final Report, Federal Department of Economic Affairs FDEA, Switzerland, 2007, 85 pages.

[21] H. Duan, M. Eugster, R. Hischier, M. Streicher-Porte, J. Li, Life cycle assessment study of a Chinese desktop personal computer, Sci. Total Environ. 407 (5) (2009) 1755–1764, doi.org/10.1016/j.scitotenv.2008.10.063.

[22] University of Michigan, Center for Sustainable Systems – Factsheets – Green IT, Pub. No. CSS09-07, 2019, 2 pages, css.umich.edu/sites/default/files/Green%20IT_CSS09-07_e2019.pdf.

[23] M.L. Socolof, J.G. Overly, L.E. Kincaid, J.R. Geibig, Desktop Computer Displays: A Life-Cycle Assessment, vol. 1, Center for Clean Products and Clean Technologies, University of Tennessee, 2001. Report prepared for the US Environmental Protection Agency, Office of Pollution Prevention and Toxics, Report EPA-744-R-01-004a, 995 pages, www.epa.gov/sites/production/files/2014-01/documents/computer_display_lca.pdf.

[24] U.S. Environmental Protection Agency, Documentation Chapters for Greenhouse Gas Emission, Energy and Economic Factors Used in the Waste Reduction Model (WARM) — Electronics Chapter, 23 pages. www.epa.gov/sites/production/files/2019-06/documents/warm_v15_electronics.pdf.

[25] J. Malmodin, D. Lundén, Å. Moberg, G. Andersson, M. Nilsson, Life cycle assessment of ICT — carbon footprint and operational electricity use from the operator, national, and subscriber perspective in Sweden, J. Ind. Ecol. 18 (6) (2014) 829—845, doi.org/10.1111/jiec.12145.

[26] E. Williams, Environmental impacts in the production of personal computers, in: R. Kuehr, E. Williams (Eds.), Chapter 3 in Computers and the Environment: Understanding and Managing Their Impacts, Kluwer Acad. Pub. (now Springer), 2003, pp. 41—72.

[27] L. Deng, C.W. Babbitt, E.D. Williams, Economic-balance hybrid LCA extended with ncertainty analysis: case study of a laptop computer, J. Cleaner Prod. 19 (11) (2011) 1198—1206, doi.org/10.1016/j.jclepro.2011.03.004.

[28] B.V. Kasulaitis, C.W. Babbitt, R. Kahhat, E. Williams, E.G. Ryen, Evolving materials, attributes, and functionality in consumer electronics: case study of laptop computers, Resour. Conserv. Recycl. 100 (2015) 1—10, doi.org/10.1016/j.resconrec.2015.03.014.

[29] R. Liu, S. Prakash, K. Schischke, L. Stobbe, State of the art in life cycle assessment of laptops and remaining challenges on the component level: the case of integrated circuits, in: M. Finkbeiner (Ed.), Towards Life Cycle Sustainability Management, Springer, 2011, pp. 501—512.

[30] J. Ciroth, J. Franze, LCA of an Ecolabeled Notebook — Consideration of Social and Environmental Impacts along the Entire Life Cycle. Report Prepared by GreenDeltaTC GmbH, Berlin for the Federal Public Planning Service Sustainable Development (FINTO) in Brussels, 2011, 409 pages, www.greendelta.com/wp-content/uploads/2017/03/LCA_laptop_final.pdf.

[31] M. Sloma, Carbon footprint of electronic devices, Proc. SPIE 8902 (2013) 890225, doi.org/10.1117/12.2030271. Electron Technology Conf. 2013, Ryn, Poland.

[32] H. André, M. Ljunggren Söderman, A. Nordelöf, Resource and environmental impacts of using second-hand laptop computers: a case study of commercial reuse, Waste Manag. 88 (2019) 268—279.

[33] J. Quariguasi-Frota-Neto, J. Bloemhof, An analysis of the eco-efficiency of remanufactured personal computers and mobile phones, Prod. Op. Manag. 21 (1) (2012) 101—114, doi.org/10.1111/j.1937-5956.2011.01234.x.

[34] R. Hischier, P. Wäger, J. Gauglhofer, Does WEEE recycling make sense from an environmental perspective? The environmental impacts of the Swiss take-back and recycling systems for waste electrical and electronic equipment (WEEE), Environ. Impact Assess. Rev. 25 (5) (2005) 525—539, doi.org/10.1016/j.eiar.2005.04.003.

[35] E. Van Eygen, S. De Meester, H.P. Tran, J. Dewulf, Resource savings by urban mining: the case of desktop and laptop computers in Belgium, Resour. Conserv. Recycl. 107 (2016) 53—64, doi.org/10.1016/j.resconrec.2015.10.032.

[36] P.A. Wäger, R. Hischier, M. Eugster, Environmental impacts of the Swiss collection and recovery systems for Waste Electrical and Electronic Equipment (WEEE): a follow-up, Sci. Total Environ. 409 (2011) 1746—1756, doi.org/10.1016/j.scitotenv.2011.01.050.

[37] T. Zink, F. Maker, R. Geyer, R. Amirtharajah, V. Akella, Comparative life cycle assessment of smartphone reuse: repurposing vs. refurbishment, Int. J. Life Cycle Assess. 19 (5) (2014) 1099−1109, doi.org/10.1007/s11367-014-0720-7.

[38] F. Suarez, A. Nozariasbmarz, D. Vashaee, M. Öztürk, Designing thermoelectric generators for self-powered wearable electronics, Energy Environ. Sci. 9 (2016) 2099−2113, doi.org/10.1039/C6EE00456C.

[39] J. Vaajakari, How Sustainable Is Wearable Technology? Data Driven Investor, Posted 17 September 2018, 2018. medium.com/datadriveninvestor/how-sustainable-is-wearable-technology-88608a932cb4.

[40] G. Katarzyna, T. Tomasz, Life cycle assessment of an inkjet printer, Polish J. Environ. Stud. 21 (5A) (2012) 95−105.

[41] Lexmark International Inc, Environmental Product Declaration − Laser Printer CS820DE, February 17, 2016, 22 pages, csr.lexmark.com/pdfs/2.23.16/101.1_Lexmark_EPD_CS820DE.pdf.

[42] M. Bozeman, V. DeYoung, W. Latko, C. Schafer, Life Cycle Assessment of a Solid Ink Printer Compared with a Color Laser Printer − Total Lifetime Energy Investment and Global Warming Impact, Xerox Corporation, 2010, 8 pages, www.office.xerox.com/latest/887WP-01U.PDF.

[43] W. Kerr, C. Ryan, Eco-efficiency gains from remanufacturing − a case study of photocopier remanufacturing at Juji Xerox Australia, J. Cleaner Prod. 9 (1) (2001) 75−81, doi.org/10.1016/S0959-6526(00)00032-9.

[44] Apple Inc, Environmental Report − Apple TV 4K, September 2017, 4 pages, www.apple.com/environment/pdf/products/appletv/Apple_TV_4K_PER_sept2017.pdf.

[45] R. Hischier, I. Baudin, LCA study of a plasma television device, Int. J. Life Cycle Assess. 15 (2010) 428−438, doi.org/10.1007/s11367.010.0169.2.

[46] Waste and Resources Action Programme (WRAP.org.uk), Environmental Assessment of Consumer Electronic Products, May 20, 2010, 24 pages, www.wrap.org.uk/sites/files/wrap/Environmental%20assessment%20of%20consumer%20electronic%20products.pdf.

[47] W.K.C. Yung, S.S. Muthu, K. Subramanian, Carbon footprint analysis of personal electronic product − induction cooker. Chap. 5 in S.S. Muthu (Ed.), Environmental Carbon Footprints − Industrial Case Studies, Butterworth-Heinemann, Elsevier, 2018, pp. 113−140, doi.org/10.1016/B978-0-12-812849-7.00005-2.

[48] S. George, D. McKay, The Environmental Impact of Music: Digital Records, CDs Analysed. The Conversation − Environment + Energy, Posting Dated 10 January 2019, 2019. theconversation.com/the-environmental-impact-of-music-digital-records-cds-analysed-108942.

[49] J. Faludi, C. Bayley, S. Bhogal, M. Iribarne, Comparing environmental impacts of additive manufacturing vs. traditional machining via life-cycle assessment, Rapid Prototyp. J. 21 (1) (2015) 14−33, doi.org/10.1108/RPJ-07-2013-0067.

[50] Y. Luo, Z. Ji, M.C. Leu, R. Caudill, Environmental Performance Analysis of Solid Freeform Fabrication Processes. Proc. 1999 IEEE International Symposium on Electronics and the Environment 1999, ISEE -1999, 1999, pp. 1−6, doi.org/10.1109/ISEE.1999.765837.

[51] J. Faludi, Z. Hu, S. Alrashed, C. Braunholz, S. Kaul, L. Kassaye, Does material choice drive sustainability of 3D printing? Int. J. Mech. Aero. Industr. Mechatron. Eng. 9 (2) (2015) 144−151. digitalcommons.dartmouth.edu/facoa/2111.

[52] I. Muñoz, C. Gazulla, A. Bala, et al., LCA and ecodesign in the toy industry, case study of a teddy bear incorporating electric and electronic components, Int. J. Life Cycle Assess. 14 (2009) 64−72, doi.org/10.1007/S11367-008-0044-6.

Chapter 8

Information and internet

Following on Chapter 7 that covered the impacts of electronic hardware, the present chapter addresses the environmental effects of *using* such hardware, the so-called Information Communications Technology (ICT), which includes internet information storage, posting and retrieval, online business dealings, online shopping, and the many uses of a smartphone. It also extends to several second-order effects, such as the green aspects of paperless communications, *e*-commerce (fewer stores, more warehouses), and travel reduction by video conferencing. One thing is certain: Digitalization causes a significant increase in electricity consumption, particularly at peak times [1]. Trends to 2030 and to 2040 predict that, in a worst-case scenario, electricity usage for ICT could contribute up to 23% of all greenhouse gas emissions by 2030 [2], and that by 2040 the footprint of smartphones may exceed that of desktops and laptops. Greenhouse gas emissions attributable to ICT can eventually amount to half of the whole transportation sector [3].

A word of caution is in order: This is an area with scant and scattered information subject to rapid changes. Therefore, the numbers provided here should be considered as indicative rather than precise.

8.1 Internet infrastructure

8.1.1 Data centers

The backbone of the world wide web (internet) is a large and growing set of servers (electronic equipment similar to computers but without keyboard and display) grouped in data centers. There can be upward of 50,000 servers in one data center [4]. Over its lifetime, the carbon footprint of a typical server has been estimated to be $6,360 \, kg \, CO_{2eq}$ when used in the United States, of which 90% ($5,960 \, kg \, CO_{2eq}$) is from its use and 7% ($471 \, kg$ of CO_{2eq}) from its manufacturing [5].

These data centers, which together form the "cloud," can handle tens of millions of queries per second [6] and consume an estimated 205 TWh (terawatt-hours) of electricity in 2018 or about 1% of global electricity consumption [7]. In the United States, the electricity consumption was estimated (circa 2018) at 2.4% of the national electricity consumption [8], with about as much carbon emissions as the aviation industry [9]. The energy consumption by data centers is due only in parts to the servers as there is also a significant overhead for operations, mostly because of the need to cool the space to

Data, Statistics, and Useful Numbers for Environmental Sustainability.
https://doi.org/10.1016/B978-0-12-822958-3.00003-0

evacuate the heat produced by the servers. It is estimated that the power usage effectiveness (PUE) of data centers, which is the ratio of the total energy demand to that of the servers only, is approximately 1.9 [10] or 2.0 [11] although Google claims that the PUE of their data centers is only 1.11 while the industry's average is 1.67 ([12] page 19). Likewise, Facebook claims that the PUE of its data centers is 1.11 ([13] page 15).

The annual energy cost of a data center per square foot is 15–40 times that of an office building [10].

The energy consumption of a data center is primarily a function of the number of operating servers it houses and cannot be directly linked to the amount of digital service being provided. So, the energy intensity of the internet, which could be defined as the energy consumed per gigabyte transmitted, cannot be calculated, at least not at the level of an individual data center. Estimates have ranged from as low as 0.0064 kWh/GB to as high as 136 kWh/GB, with a possible world average of 0.39 kWh/GB [14]. Other authors estimate that the average operating energy of the internet is 1.59 kWh/GB [15] or lies between 2.17 and 3.61 kWh/GB [16]. The number is also a moving target as servers and data centers that house them become increasingly more energy efficient over time, with a reduction of energy use by a factor of 50 over 11 years, corresponding to the rate of decrease of 30% per year [14].

It is estimated that the ratio of servers to desktop computers in business networks is 1:30 ([17] page 836).

8.1.2 Internet of Things

The Internet of Things, or IoT in short, is the network of sensors embedded in our smartphones, cars, homes, buildings, public spaces, and infrastructure in general with the ability to communicate data in real time with appropriate software in the "cloud" that in turns enables decision making and management with minimal human control. In our houses, it takes the form of smart thermostats and smart appliances (ex. refrigerators that formulate shopping lists), and along our streets in the form of traffic sensors controlling traffic lights to minimize bottlenecks. As such, IoT holds the promise of optimization and error avoidance, and consequently energy efficiency and lowered environmental burdens. Yet, like all new technologies, it creates impacts of its own.

The hardware supporting the IoT is a vast array of sensors, WiFi communication devices, and connected servers. The impacts of these electronic devices are known to consist of their embodied energy (especially in electronic chips), their electricity consumption during use, and inadequate recycling of e-waste (see Sections 7.3.4 and 11.3). The indirect impacts can be both negative and positive. On the negative side, one expects an increase in added electricity consumption to communicate and store the abundant data as well as increased electronics waste. On the positive side, one expects increased energy efficiency, optimized logistics, and safer industrial operations, each of which has beneficial environmental implications. Google estimates that its Nest smart thermostat has proven energy savings of 10–12% for heating and 15% for cooling ([12] page 49).

On the most hopeful note, it has been suggested that IoT can help attain the Sustainable Development Goals set by the United Nations as enumerated in its the 2030 Agenda for Sustainable Development, which includes clean water for all, climate change mitigation, and industry efficiency, among others [18]. On the ecosystem side, IoT can improve resource preservation and endangered species protection [19]. In balance, IoT should be more beneficial than detrimental, but the net effect has yet to be quantified.

8.1.3 Optical fibers

If servers are the brain of the internet, optical fibers are its arteries. It has been estimated that the production of optical fibers causes an emission of 4.81 g of CO_{2eq} per meter of fiber, equivalent to 72.9 kg of CO_{2eq} per kg of fiber [20]. This source does not mention the energy spent in manufacturing, but it can be reversed engineered. Using the ratio used in the analysis of 0.614 kg of CO_{2eq} per kWh of electricity used, it is estimated that the manufacturing of optical fibers (excluding procurement of raw materials) is 28.2 kJ/m on a length basis and 427 kJ/kg on a mass basis.

According to a major manufacturer [21], optical fibers do not wear over time and thus have no maximum theoretical lifetime. Operators install fiber cables with the expectation of a minimum of 25 years of service.

The energy consumption per bit of data transmitted through optical IP networks has been estimated at 75 μJ at low access rates and dropping to 2−4 μJ at an access rate of 100 Mb/s [22]. This amounts to 0.18 kWh/GB at low transmission rates and dropping to 0.005−0.01 kWh/GB at high rates. A separate study pegs the consumption at 0.08 kWh/GB, with the concomitant carbon footprint of 0.048 kg CO_{2eq}/GB (using the very low carbon footprint of electricity in Sweden: 0.06 kg CO_{2eq}/kWh, whereas the global electricity mix suggests 0.6 kg CO_{2eq}/kWh) ([17] Footnote below Table 1).

A single optical fiber can carry 15−20 mW of power for approximately 1−10 Gb/s. Signal attenuation in an optical fiber is estimated at 2 dB per km, equivalent to a 5% reduction in power for every km.

International submarine cables consume 0.02 kWh/GB ([17] Table 3).

Generally, 10−20 individual fibers are bundled into a cable, and 6−10 of these cables are assembled into a thicker cable, which is then enclosed in a protective cover. These optical fiber cables can be disassembled at the end of their life, and their materials recycled [23].

8.2 Communications

Most of our communications nowadays are conveyed digitally, and there is no value in trying to estimate the number of terabytes that are communicated every day since this number is growing rapidly, and the energy consumed per gigabyte transmitted by mobile devices as decreased exponentially from 12.34 kWh/GB in 2010 to 0.30 kWh/GB in 2017 ([24] Fig. 3). Attempts can be made, however, to assign an environmental impact to specific activities.

8.2.1 Electronic messaging

The carbon footprint of sending a text message from one mobile phone to another can be estimated by the time it takes the sender to type the words on their screen plus the time it takes the recipient to read those words at the other end. If the total time amounts to 1 minute on mobile phones that consume 1 W of power when in use[1], we estimate the electricity consumption at 1/60th of a Wh. Based on the ratio of 0.45 kg (0.99 lbs) of CO_2 per kWh for the electricity mix in the United States (quoted in Section 4.3.4.), we obtain that the carbon footprint of a simple text exchange is 0.0075 g (on the order of the 0.014 g estimated in Ref. [25] page 16).

A nonspam email that is hand-typed by the sender, has no attachment, and is fully read by the recipient is estimated to be responsible for a 4 g emission of CO_2, rising to 50 g if there is a photo or large text attachment, whereas a mass-produced spam message that is not read causes a much smaller emission of 0.3 g ([25], page 20). This compares very favorably to sending a letter in the mail: from 140 g of CO_{2eq} for a letter written on recycled paper and recycled afterward to 200 g of CO_{2eq} for a letter on virgin paper that is not recycled afterward ([25] page 45). Depending on the volume sent by the individual, emailing may result in 1.6 kg (3.5 lb) CO_{2eq} per day, or 135 kg (298 lb) CO_{2eq} per year [26].

8.2.2 Smartphone use

A 1-minute phone call has about the same carbon footprint as a text message, 0.14 g CO_{2eq} [26].

Excluding the use of the network and the burden of data centers, a typical smartphone with a lifetime of 3 years causes about 19 kg of CO_{2eq} emissions per year, but when network and data centers are added to the tally at the rate of 43 kg CO_{2eq} per year, the number rises to 62 kg CO_{2eq} per year [27]. Thus, of the various segments of the telecommunication industry, including material extraction, manufacture, operation/use, and end-of-life management, the network is the most impactful element with almost 50% of the total impact.

Another source estimates the typical carbon footprint of a mobile phone user at 25 kg of CO_{2eq}/year, noting that it is equivalent to driving an average European car on the road for about 1 hour [28].

Smartphone use is associated with a few wasteful practices. Firstly, convenience prompts the user to receive and send an abundance of data (hundreds of megabytes per week), some of which is barely considered. Then, social networks that most often connect local people have the data routed through distant data centers. Users typically use only a handful of apps while their phone carries dozens of them, each of which is updated regularly. It has also

---

1. The average power of a mobile phone is 0.3 W ([17] Table 1), but this includes idling.

been noted that peak data consumption tends to coincide with peak electricity demand when the electricity is less clean [29].

8.2.3 Cellular telephone towers

The infrastructure supporting mobile phone communications consumes 10 times more energy than the device itself [1]. Also, using a smartphone over a mobile network is almost twice as energy-intensive as using it over a fixed (Wi-Fi) network, 35 Wh versus 20 Wh per 100 MB [29].

A typical cellular telephone tower consumes 4–6 kW, less in urban areas and more outside of urban areas. The geographical density of towers amounts to 0.4 kW per km^2 (1.0 kW/mi^2) of the covered area [30]. At 5 kW per tower in average, this translates to 1 tower serving an area of 12.5 km^2 (about 5 mi^2). In Sweden, there are 27–33 mobile devices per km^2 (70–85 devices per mi^2) ([17] page 837), for an estimate of 340–410 mobile devices served per tower. This number, of course, is very coarse as it is very much dependent on population density and also subject to change over time as networks progress from 4G to 5G.

Some people worry about human exposure to radiofrequency fields emitted by cell phone towers. Although the Federal Communications Commission of the United States limits their radiated power to 500 watts per channel, the majority of cellular towers in urban and suburban areas operate at 100 watts per channel or less [31]. This is well below the exposure limits recommended for radiofrequency and microwave safety standards [31].

8.3 Internet usage and digital consumption

In 2018, the world generated 33 zettabytes (=33 × 2^{70} bytes) of data, the equivalent of 7 trillion DVDs ([32] page 5). Google states that it offers 20 petabytes (=20 × 2^{50} bytes) of geospatial data ([12] page 15).

8.3.1 Web searches

According to Google (as reported in Ref. [25] pages 17 and 200), a web search at their end consumes 0.3 Wh (about 1 kJ) causing a carbon footprint of 0.2 g. After the inclusion of the impact of the device used to run the search, which might be a 20 W laptop used for 30 s, the carbon footprint rises to 0.7 g CO_{2eq} per search. More broadly, and still according to Google's own data, the average user of its services who performs 25 searches per day, uses a gmail account, views 60 minutes of videos on YouTube, and accesses some of its other services produces less than 8 g CO_{2eq} per day ([12] page 62).

8.3.2 Social media

The Facebook company calculates that the average person on its platform contributed in 2019 to 0.10 kg CO_{2eq}, which is less than the carbon footprint of making one cup of coffee ([13] page 6). While this may sound benign, one should pay attention to the indirect effects. For example, the social pressure to post on the platform ever more fascinating photos of self may prompt a person or family to spend their vacation at more exotic places or to engage in more extreme activities, which often have higher environmental impacts. On the other hand, the Facebook platform facilitates environmental activism and the dissemination of good environmental practices.

The Twitter company does not appear to report on its environmental impacts.

8.3.3 Consumption of digital information

Nowadays, a college textbook may come in the form of either a physical book or a digital file, and the question arises as to which form has lower impacts. The pertinent elements for each type of book are tabulated in the following, with the estimated use time of the book corresponding to one academic semester.

Physical vs. digital textbooks: Assumptions					
Physical book			**Digital book**		
Kraft paper (pages)		3.2 lb	Hours spent reading book per week		3 hr
Cardboard—bleached (cover)		0.8 lb	Weeks in the term		16 weeks
Printing	offset printing	3.2 lb	Fraction of lifetime laptop use		1%
	rotogravure	0.8 lb	Laptop manufacturing	laptop itself	1 item
Transport		4.0 lb, 3,000 mi		adapter	1 item
Controlled incineration	paper	3.2 lb	Transport		lb, 3,000 mi
	cardboard	0.8 lb	Laptop's use		40 W
			Controlled landfill—electronics		5.0 lb

Source: Authors' own.

Transportation amounts to 6.00 ton-miles for the printed book and 7.50 ton-miles for the digitized book, whereas the electricity consumption for the digitized book is 1.92 kWh. Using Okala Impact Factors ([33] Chapter 11), the environmental scores can be tallied for each type of book as tabulated in the following.

Physical vs. digital textbooks: Comparative impacts

Book in print				Book in digital form				
Material/process	Amount	Impact factor	Net	Material/process	Amount	Impact factor	% use	Net
Kraft paper	3.2 lb	1.11/lb	3.55	Laptop	1 item	770/item	0.5%	3.85
Bleached cardboard	0.8 lb	0.84/lb	0.67	Adapter	1 item	41/item	0.5%	0.21
Offset printing	3.2 lb	1.7/lb	5.44	Electricity	1.92 kWh	0.92/kWh	100%	1.77
Rotogravure printing	0.8 lb	1.9/lb	1.52	—	—	—	—	—
Transport	6 ton-mi	0.31/ton-mi	1.86	Transport	7.5 ton-mi	0.31/ton-mi	0.5%	0.012
Incineration—paper	3.2 lb	0.012/lb	0.0384	—	—	—	—	—
Incineration—cardboard	0.8 lb	0.012/lb	0.0096	Disposal—electronics	5.0 lb	3.6/lb	0.5%	0.090
—	—	Total:	13.1			Total:		5.9

Source: Authors' own calculations using Okala Impact Factors [33].

The conclusion is that the book in digital form is preferable to the book in print, but it should be noted that this conclusion is crucially depending on the assumption that 0.5% of the laptop lifetime is dedicated to reading the digitized book. The conclusion reverses if more than 1.14% of the laptop's lifetime is used for the digitized book.

Below is a comparative life-cycle assessment of three ways of getting the daily news (incl. weather forecast), by viewing news on a television set, by reading an online newspaper, or by reading a paper print of the newspaper [34]. Assumptions are made about the amount of the printed newspaper that is actually read by the reader (43%) and the footprint of the electricity mix. "Dirty" electricity can make online reading worse than reading a printed newspaper. Viewing news on television is slightly less impacting than reading news online because television sets generally cause less impact during the manufacture stage than computers.

Comparative impacts of getting the news				
Life-cycle stage	Daily news on television	Daily news from internet newspaper	Daily news from thin newspaper (ex. tabloid)	Daily news from major newspaper
Production of hardware or paper[1]	4.0	8.0	32.1	40.0
Transportation	–	–	1.6	10.0
Distribution	–	–	2.5	6.0
Viewing or online reading	6.0	9.5	–	–
Disposal	–	–	1.6	3.2
Recycling credit	–	–	−3.6	−10.0
Total	10.0	17.5	34.2	49.2

Units are eco-points of an environmental scarcity method and are used for comparison only.
[1] Printing has a negligible impact next to paper production.
Source: [34] Fig. 5.

For a single news item, 180 seconds of television watching, 90 seconds of online reading, and the reading of 250 cm^2 of print are equally impacting [34].

In terms of carbon footprint, it is estimated that reading news online at the rate of 1 hr per week on a 50-Watt laptop causes about as much carbon emissions as a weekly news magazine like the *Guardian Weekly* ([25] page50).

The same question arises in the context of music: Is it better to download a music file than to purchase a compact disk (CD)? The answer is yes. According to Ref. [35], a CD incurs up to 53 MJ (=15 kWh) of energy and 3.2 kg CO_{2eq}

whereas the download of the digital file of the same album, with no CD burning afterward, only 7 MJ ($=2$ kWh) of energy and 0.4 kg CO_{2eq} of carbon. A separate study quotes 1.0 kWh for the download of a 700 MB music file [15].

For music, a third option is to listen online each time and never store the digital file on one's device, an option called "streaming." An answer to the question of whether it is better to stream the music than to own the compact disk was given by *The Conversation*: "If you only listen to a track a couple of times, then streaming is the best option. If you listen repeatedly, a physical copy is best; streaming an album over the internet more than 27 times will likely use more energy than it takes to produce and manufacture the same CD." (Source: [36], which appears to have obtained its information from Ref. [37]).

Similarly, video streaming, which by some estimate accounts for 60% of the world's internet traffic [26], has a significant impact. An hour of video weekly consumes annually more electricity in the networks than two new refrigerators use in a year ([38] page 3).

Assuming that 2 kWh of electricity is being consumed upstream per gigabyte being streamed, one estimates that streaming a high-definition movie at a rate of 3.0 GB/hr is tantamount to consuming a power of 6 kW and a 4K-high-definition movie at a rate of 7.0 GB/hr to consuming 14 kW [39], far more than the wattage of the plasma screen on which the movie may be viewed (20 W + 203 W/m^2 according to Ref. [16]).

The British Broadcasting Corporation estimates that viewing pornography online may account for as much as one-third of the world's video streaming traffic. Another third includes the streaming of movies on platforms such as Amazon Prime and Netflix, while watching YouTube and video clips on social media accounts for the remaining third. Each third causes a carbon footprint about as large as that of Belgium, a country of 11.7 million people [26].

In its 2019 report entitled Environmental Social Governance, Netflix boasted to have 167 million subscribers while it acknowledged being responsible for a direct electricity consumption of 94,000 MWh and an additional indirect[2] electricity consumption of 357,000 MWh ([40] page 2). This amounts to an annual electricity consumption of 2.70 kWh per subscriber, with concomitant carbon footprint of 2.06 kg (4.53 lbs) of CO_{2eq} per subscriber per year.

Another question is whether downloading video games is preferable to producing and selling disks. As for all similar questions, the answer is: It depends. For an average 8.80 GB game, the carbon footprint of the download can vary from 2.26 to 7.89 kg CO_{2eq} whereas the several steps of disk production from materials procurement to distribution and retail account for an estimated 1.07 kg CO_{2eq}, making the 101-g physical disk be the cleaner option, but the conclusion is reversed if the file to be downloaded falls below

---

2. Electricity consumed on behalf of the company by others that provide servers and distribution channels.

1.3 GB. Whatever the case may be, the upstream impact pales in comparison with the footprint of playing the game, pegged at 19.5 kg CO_{2eq} for 232 hours of play on a 137-W console [41].

8.3.4 Videoconferencing

Videoconferencing has definite advantages over in-person meetings for the evident reason that it substitutes relatively minor electricity impacts for the substantial impacts of travel. The impact reduction depends on many factors including the number of people in the meeting, where and how they would have traveled, and details such as the duration of the videoconference and the nature and number of electronic peripherals being used. For a typical scenario, the reduction in environmental impacts is around 90% [42].

The following table lists the power requirement of the typical components used in videoconferencing. The carbon footprints can be easily estimated using the footprint of a unit of electricity, such as of 0.45 kg (0.99 lbs) of CO_{2eq} per kWh for the electricity mix in the United States (quoted in Section 4.3.4.) or 0.614 kg (1.35 lbs) of CO_{2eq} per kWh in Europe.

Power requirements in videoconferencing		
Electronic component		**Power consumption**
Computer	Desktop	150 W
	Laptop	40 W
Display	Plasma screen	20 W + 203 W/m^2
	Liquid-crystal display or LED	20 W + 172 W/m^2
	Projector	135 W
Peripherals	Camera	9.5 W
	Speakers	4.1 W
	Microphone	2.5 W
Local area network (LAN)		20 W

Source: [16] Table 1.

Using these numbers as well as numbers for various means of travel, a study [16] estimated that a hypothetical 5-hour meeting held by video conferencing between four participants (two local, one being 1,000 km away and on being 5,000 km away) avoids around 20 GJ in transportation (depending on the mode of transport) in return for 0.025 GJ in electricity. When taking into account the life-cycle impact of all electronic devices involved in proportion to their use for the event, the energy footprint of the videoconference rises to 1.3 GB.

For a meeting between two people, videoconferencing is less impacting than the in-person meeting as long as travel would have been more than 15 km (9.3 mi). If the impacts of the manufacturing of the electronic devices are included in the analysis, the breakeven distance rises to 20 km (12 mi). For a distance of 1,000 km (620 mi), a 4-hour videoconference between two people needs only 1% of the primary energy if travel would have been by car or plane, or 5% for travel by train [42].

On the downside, the same authors also note that videoconferences tend to take slightly more time than the in-person meetings they replace and may be more impacting than in-person meetings when little to no travel would have been involved (ex. a meeting of persons in the same location) [16]. There can also be a so-called rebound effect: Time saved from not traveling is not spent idling, and the impact of the replacement activity should be subtracted from the savings of the videoconference [42].

8.3.5 e-Commerce

A revolutionary activity enabled by the internet is on-line purchasing of goods, the so-called *e*-commerce. This activity has created a major, yet still partial, shift from in-store purchasing to on-line ordering, with a vast series of consequences, each with its own environmental impact. Physical stores in urban areas with air conditioning, bright lights, parking lots and contributing to traffic are being replaced by dimly lit and highly robotized warehouses near major roadways; individualized vehicle travel is giving way to freight hauling by trucks and the building of airports dedicated exclusively to freight; fancy packaging turns into the digital presentation of the product followed by plain cardboard packaging; new companies are selling craft, vintage, refurbished, or specialty items; etc. [43,44]. Whether this major shift has a net positive or negative effect on the planet is not yet known and may never be, for there are too many ramifications, especially when one considers indirect effects. For example, does shopping at the convenience of a click lead to more shopping? If people save time, how do they spend their free time? If people save money, what do they do with their savings? How prevalent is the practice of visiting a physical store only to examine a product and ask questions before buying the same online later? Verdicts can only be reached in specific cases.

An early study compared the in-store and on-line ways of purchasing a desktop computer and found that on-line purchasing could result in a negative impact on the environment of about 10% of the full lifecycle of the product. This study, however, cautioned that if the *e*-retailer uses the customer data to optimize its supply chain, the conclusion could be reversed [45].

Another early study is that of the purchase of a bestseller book, with the possibility of the store returning unsold books back to the publisher. The following table compares the energy and carbon footprints per book for each mode of purchasing, and the results suggest that on-line purchasing has comparable impacts to those of traditional retailing depending on the percentage of books that the retailer returns unsold to the publisher [46].

Comparative impacts of different ways of purchasing a bestseller book			
Method of purchase	Energy (MJ)	Air pollutants (kg)	Greenhouse gases (kg CO_{2eq})
In-store, no returns[1]	98	0.2	6
In store, 35% returns[1]	115	0.22	7
On-line	105	0.12	6.9

[1]Based on a 5-mi (8 km) errand and on the store returning unsold books (0 or 35%) to the publisher.
Source: [46].

In a densely populated country, such as Japan, distances are not only shorter, but transport may also be by train instead of truck or plane. When this is the case, the numbers (tabulated in the following) change significantly, showing that energy consumption is much lower in a densely populated area and that traditional in-store purchasing is clearly the greener option then.

In-store versus on-line purchase of an item in two different locations			
Energy footprint (MJ)	United States		Tokyo, Japan
In-store purchase	60−86		1.35
On-line purchase	91	104	5
	Transport by truck	Transport by plane	Transport by train

Source: [47].

A study [43] comparing in-store and on-line purchasing of a flash drive (about 1 lb with packaging) from buy.com showed that *e*-commerce has lower energy and carbon footprints (2,200 g CO_{2eq}/item) than traditional retail (2,800 g CO_{2eq}/item), a reduction of 21%. The major savings come mostly from the avoidance of the customer's travel to the store, which for traditional retail accounts for 1,820 g CO_{2eq}/item (=65%) of the footprint, but these are partly offset by the new impacts of extra packaging (484 g CO_{2eq}/item, 22%) and last-mile delivery (704 g CO_{2eq}/item, 32%). When the purchase is air-freighted to the customer, the footprint of *e*-commerce rises to nearly 2,500 g CO_{2eq}/item. The impacts of warehousing are about the same in each case (around 600 g CO_{2eq}/item). Since the product considered amounts to about 1 lb with its packaging, all preceding numbers on a per-item basis may be interpreted as on a per-pound basis for similar products. These findings were corroborated by Ref. [48].

If the product purchased online is a digital file, the method of delivery is a simple download, at a cost of 1.59 kWh/GB [15] with an accompanying carbon footprint of 0.8 kg CO_{2eq}/GB.

The carbon footprint of the so-called "last mile" varies from an average of 181 g CO_{2eq} for an online package delivered by a van that delivers a total of 120 packages in a 50-mile circuit to 4,274 g CO_{2eq} for a 12.8-mile roundtrip to the store by personal car for a single item. A shopper riding a bus for 8.8 miles and in the company of 29 other passengers to buy a total of five items is responsible for a carbon footprint of only 78 g CO_{2eq} per item [49]. Extending these footprints to arbitrary numbers of miles (or km) and of items carried on board the delivery van, one obtains the following formulas:

$$\text{Online order delivered by van}: \frac{433 \times m_{van}}{n_{van}} \frac{\text{g } CO_{2eq}}{\text{item}} = \frac{269 \times k_{van}}{n_{van}} \frac{\text{g } CO_{2eq}}{\text{item}}$$

$$\text{In-store purchase by car}: \frac{334 \times m_{errand}}{n_{purchases}} \frac{\text{g } CO_{2eq}}{\text{item}} = \frac{207.5 \times k_{errand}}{n_{purchases}} \frac{\text{g } CO_{2eq}}{\text{item}}$$

$$\text{In-store purchase by bus}: \frac{44.3 \times m_{errand}}{n_{purchases}} \frac{\text{g } CO_{2eq}}{\text{item}} = \frac{27.5 \times k_{errand}}{n_{purchases}} \frac{\text{g } CO_{2eq}}{\text{item}},$$

for a delivery van that delivers a total of n_{van} packages along a circuit of m_{van} miles (k_{van} kilometers) versus a person driving a personal car or riding a bus for m_{errand} miles (k_{errand} kilometers) on a shopping errand for $n_{purchases}$ items.

Proceeding from a survey of shoppers instead of performing a life-cycle analysis, Dutch investigators [48] considered how online purchasing affects the number of trips and travel distances. The following table gives the number of trips and kilometers added (positive) or avoided (negative) for two types of online purchases, those from retailers (B2C, Business-to-Consumer) and from individuals through used-good platforms (C2C, Consumer-to-Consumer). Since the absolute numbers given in the study are particular to time (2006) and place (Netherlands), the following tabulated numbers refer to an arbitrary 100 personal trips to a store avoided by B2C online purchases. Compared to in-store purchases, B2C online purchases cause a reduction in total driving whereas C2C online purchases causes a slight increase. The conclusion is that for every 100 trips to the store avoided by online purchasing, 1,141 km of personal travel are avoided and 366 km of freight travel are added.

Travel impacts when switching from in-store to on-line purchasing						
e-Commerce type	Trips avoided (−) or added (+)			Kilometers avoided (−) or added (+)		
	Personal trips	Freight trips	Net	Personal distance	Freight distance	Net
B2C	−100	+267.8	+167.8	−1,352	+304	−1,048
C2C	+11.6	+54.9	+66.5	+211	+62	+273
Both types	−88.4	+322.7	+234.3	−1,141	+366	−775
Source: [48].						

8.3.6 Online advertising

The internet offers powerful avenues for advertising, and several platform companies have placed advertisement revenues at the core of their business models. Also, e-commerce could not proceed without some online advertising. As a result, a significant fraction of internet traffic nowadays consists of advertisements, almost all with photos and many with video clips. A question is what is the fraction of energy (and thus also of carbon footprint) attributable to online ads. The answer depends on two factors, through which channel advertisements are being delivered and which device is being used to view the ads. The following table provides some estimates. These percentages may be useful when estimating the additional burden caused by advertising on specific internet activities. Since all energy is in the form of electricity, the percentages listed in the following are also those of carbon footprints.

Energy burden of advertising on various platforms						
Type of internet traffic	Estimated fraction of time that is advertising	Fraction of total energy attributable to advertising				
		Desktop computer	Laptop	Tablet	Smartphone	All devices
Web, email, and data	25–75%	26.0%	7.88%	3.25%	8.69%	45.8%
File sharing	1–19%	5.04%	1.54%	0.65%	0.02%	7.25%
Video viewing	2–25%	29.4	8.85%	3.74%	3.90%	45.9%
Video gaming	1–19%	0.73%	0.24%	0.08%	0.00%	1.05%
All traffic combined	30%[1]	61.2%	18.5%	7.7%	12.6%	100%

[1]*Energy-weighted average.*
Source: [50].

8.3.7 Platform economy

The enabling of commercial transactions by the internet has spawned a new economic sector called the platform economy [51]. A platform is an intermediate digital agent that links sellers with customers: Sellers offer their goods on services on the platform, while customers reach out to the platform to purchase these goods and services. Like all other types of business, a platform impacts the physical environment both positively and negatively.

One claim is that platforms inject "intelligence" in the conduct of business, which naturally leads to optimization and, in its wake, conservation of resources ([51] page 272). This has yet to be quantified. Other benefits and negative impacts are sector specific.

8.3.7.1 Ride-hailing companies

At first thought, ride-hailing companies such as Uber and Lyft, may be perceived as environmentally friendly as one car can serve a large number of travelers daily, especially when allowing multiple passengers per ride ("pooling"). They have the potential to free urban areas from parking spaces, especially, at airports. Car sharing should also lower the total number of cars owned by people, a form of dematerialization ([51] page 62).

However, the findings point to some significant negative effects. The personal convenience of ride-hailing has caused a decline in the use of public transportation and a concomitant increase in street-level traffic congestion [52]. They cause "deadheading," that is, miles being driven without passengers between successive customers [53]. The Union of Concerned Scientists estimates that because of deadheading the average ride-hailing trip produces 69% more carbon emissions than the trip it replaces and 47% more emissions than a private car [53], as shown in the following table.

Carbon footprint of ride hailing	
Mode of transportation	Carbon footprint of 1-mile ride per passenger
Private car	464 g CO_{2eq}
Ride-hailing, single passenger	683 g CO_{2eq}
Ride-hailing, pooling	455 g CO_{2eq}
Ride-hailing, single passenger in electric car	220 g CO_{2eq}
Ride-hailing, pooling in electric car	146 g CO_{2eq}
Public transit	103 g CO_{2eq}

Source: [53].

8.3.7.2 Craft and vintage goods

Platforms such as eBay and Etsy, among many others, have enlarged very significantly the market for used, craft, and collectible items. Among other indirect effects, they have spawned a new cottage industry wherein people with basic skills and time on their hands purchase broken items, repair them, and then resell them on the platform for a profit. As it constitutes a form of remanufacturing that would otherwise be uneconomical, this activity is environmentally beneficial. On the other hand, the selling of low-value used goods causes the need for packaging and long-distance shipment of items that could have otherwise created less impact in a landfill.

Although most of these companies issue annual environmental reports, their focus is on lowering their corporate carbon footprint by increasing the energy efficiency of their infrastructure, particularly their data servers, and by purchasing renewable forms of electricity. They do not document the environmental impacts of the economic activity they generate, such as their impact on packaging and freight transport [54], and how much they may contribute to the circular economy. The exception to the rule appears to be Etsy, which aims to offset 100% of the carbon footprint of its customers' shipments by protecting forests and investing in renewable energy [55]. However, the company does not provide specific numbers.

Sources

[1] J. Morley, K. Widdicks, M. Hazas, Digitalisation, energy and data demand: the impact of Internet traffic on overall and peak electricity consumption, Energy Res. Social Sci. 38 (2018) 128–137, doi.org/10.1016/j.erss.2018.01.018.

[2] A.S.G. Andrae, T. Edler, On global electricity usage of communication technology: trends to 2030, Challenges 6 (2015) 117–157, doi.org/10.3390/challe6010117.

[3] L. Belkhir, A. Elmeligi, Assessing ICT global emissions footprint: trends to 2040 & recommendations, J. Cleaner Prod. 177 (2018) 448–463, doi.org/10.1016/j.jclepro. 2017.12.239.

[4] P. Johnson, With the Public Clouds of Amazon, Microsoft and Google, Big Data Is the Proverbial Big deal, Forbes online magazine, 2017 posted 15 June 2017, www.forbes.com/ sites/johnsonpierr/2017/06/15/with-the-public-clouds-of-amazon-microsoft-and-google-big-data-is-the-proverbial-big-deal/#17647a2d2ac3.

[5] M. Stutz, Carbon Footprint of a Typical Rack Server from Dell, Company Communication, Dell Corporation, 2011, 4 pages, i.dell.com/sites/content/corporate/corp-comm/en/documents/dell-server-carbon-footprint-whitepaper.pdf.

[6] Google Data Center FAQ, Everything you ever wanted to know (and didn't) about Google data centers but were afraid to ask. Posting on Data Center Knowledge, March 17, 2017. www.datacenterknowledge.com/archives/2017/03/16/google-data-center-faq.

[7] E. Masanet, A. Shehabi, N. Lei, S. Smith, J. Koomey, Recalibrating global data center energy-use estimates, Science 367 (6481) (2020) 984–986, doi.org/10.1126/science.aba3758.

[8] University of Michigan, Center for Sustainable Systems — Factsheets — Green IT, Pub. No. CSS09-07, 2019, 2 pages, css.umich.edu/sites/default/files/Green%20IT_CSS09-07_e2019.pdf.

[9] I. Ivanova, Coronavirus Is Pushing Work Online. Is that Good for the Planet? CBS NewsMoneyWatch, 2020, 22 April 2020, www.cbsnews.com/news/online-work-cloud-computing-telecommuting-carbon-impact/.

[10] S. Greenberg, E. Mills, B. Tschudi, P. Rumsey, Best Practices for Data Centers: Lessons Learned from Benchmarking 22 Data Centers. ACEEE Summer Study on Energy Efficiency in Buildings, 2006, 13 pages.

[11] W. Vereecken, W. Van Heddeghem, D. Colle, M. Pickavet, P. Demeester, Overall ICT Footprint and green Communication Technologies. 4th International Symposium on

Communications, Control and Signal Processing (ISCCSP 2010), 2010, doi.org/10.1109/ISCCSP.2010.5463327.

[12] Google, Environmental Report 2019, 66 pages. services.google.com/fh/files/misc/google_2019-environmental-report.pdf.

[13] Facebook, Sustainability Report 2019, 2019, 17 pages, sustainability.fb.com/wp-content/uploads/2020/07/Sustainability_Report_2019-2.pdf.

[14] V.C. Coroama, D. Schien, C. Preist, L.M. Hilty, The energy intensity of the internet: home and access networks, in: L.M. Hilty, A. Bernard (Eds.), ICT Innovations for Sustainability, Springer, 2015, pp. 137–155, doi.org/10.1007/978-3-319-09228-7_8.

[15] K. Craig-Wood, P.J. Krause, Towards the estimation of the energy cost of Internet mediated transactions, Tech. Rep. Energy Effi. Comput. Spec. Interest Group (2013). April 2013, 14 pages.

[16] D. Ong, T. Moors, V. Sivaraman, Comparison of the energy, carbon and time costs of videoconferencing and in-person meetings, Comput. Commun. 50 (2014) 86–94, doi.org/10.1016/j.comcom.2014.02.009.

[17] J. Malmodin, D. Lundén, Å. Moberg, G. Andersson, M. Nilsson, Life cycle assessment of ICT − carbon footprint and operational electricity use from the operator, national, and subscriber perspective in Sweden, J. Ind. Ecol. 18 (6) (2014) 829–845, doi.org/10.1111/jiec.12145.

[18] Banco Bilbao Vizcaya Argentaria, The Internet of things and its impact on sustainability. www.bbva.com/en/the-internet-of-things-and-its-impact-on-sustainability/.

[19] Stanford University, Management Science and Engineering − Blog 238: Leading Trends in Information Technology − entry posted by spak7 on, July 27, 2018. mse238blog.stanford.edu/2018/07/spak7/how-the-internet-of-things-affects-the-environment/.

[20] S. Inakollu, R. Morin, R. Keefe, Carbon footprint estimation in fiber optics industry: A case study of OFS Fitel, LLC, Sustainability 9 (5) (2017) 865–880, doi.org/10.3390/su9050865. Also available from: mdpi.com/.

[21] Corning, Inc, Frequently Asked Questions on Fiber Reliability, April 2016, 2 pages, www.corning.com/media/worldwide/coc/documents/Fiber/RC-%20White%20Papers/WP5082%203-31-2016.pdf.

[22] J. Baliga, R. Ayre, K. Hinton, W.V. Sorin, R.S. Tucker, Energy consumption in optical IT networks, J. Lightwave Technol. 27 (13) (2009) 2391–2403, doi.org/10.1109/JLT.2008.2010142.

[23] E. Wright, A. Azapagic, G. Stevens, W. Mellor, R. Clift, Improving recyclability by design: A case study of fibre optic cable, Resour. Conserv. Recycl. 44 (1) (2004) 37–50, doi.org/10.1016/j.resconrec.2004.09.005.

[24] H. Pihkola, M. Hongisto, O. Apilo, M. Lasanen, Evaluating the energy consumption of mobile data transfer − from technology development to consumer behavior and life cycle thinking, Sustainability 10 (2494) (2018), doi.org/10.3390/su10072494, 16 pages.

[25] M. Berners-Lee, How Bad Are Bananas? The Carbon Footprint of Everything, Greystone Books, D&M Publishers, 2011, 232 pages.

[26] S. Griffiths, Why Your Internet Habits Are Not as Clean as You Think. Smart Guide to Climate Change, British Broadcasting Corporation (BBC), 2020 posted 5 March 2020, www.bbc.com/future/article/20200305-why-your-internet-habits-are-not-as-clean-as-you-think.

[27] M. Ercan, J. Malmodin, P. Bergmark, E. Kimfalk, E. Nilsson, Life cycle assessment of a smartphone, in: Advances in Computer Science Research, Proceedings of 4th Int. Conf. on ICT for Sustainability (ICT4S 2016), 2016, pp. 124−133, doi.org/10.2991/ict4s-16.2016.15, pdf available from: www.ericsson.com/en/reports-and-papers/research-papers/life-cycle-assessment-of-a-smartphone.

[28] GSMA − Mobile Technology, Health and the Environment − Environmental Impact of Mobile Communications Networks, May 8, 2009, 2 pages, www.gsma.com/publicpolicy/wp-content/uploads/2012/04/environmobilenetworks.pdf.

[29] C. Lord, M. Hazas, A.K. Clear, O. Bates, R. Whittam, J. Morley, A. Friday, Demand in my pocket: Mobile devices and the data connectivity marshalled in support of everyday practice, Proc. 2015 ACM Conf. Hum. Factors Comput. Syst. Seoul (2015) 2729−2738, doi.org/10.1145/2702123.2702162.

[30] B. Kurt, Answer to the question "What's the power consumption of the average cell phone tower/site?" on Quora, March 4, 2018. www.quora.com/Whats-the-power-consumption-of-the-average-cell-phone-tower-site.

[31] U.S. Federal, Communications Commission (FCC) − Consumer Guides - Human Exposure to Radio Frequency Fields: Guidelines for Cellular Antenna Sites, updated, October 15, 2019, 2 pages, www.fcc.gov/consumers/guides/human-exposure-radio-frequency-fields-guidelines-cellular-and-pcs-sites.

[32] Technology Quarterly, Artificial intelligence and its limits − Steeper than expected. The Economist. Insert Suppl., 13−19 June 2020 issue, 12 pages.

[33] P. White, L.S. Pierre, S. Belletire, Okala Practitioner −Integrating Ecological Design, Okala Team, Phoenix Arizona, 2013, 82 pages.

[34] I. Reichart, The environmental impact of getting the news: a comparison of on-line, television, and newspaper information delivery, J. Ind. Ecol. 6 (3−4) (2002) 185−200, doi.org/10.1162/108819802766269593.

[35] C.L. Weber, J.G. Koomey, H.S. Matthews, The energy and climate change implications of different music delivery methods, J. Ind. Ecol. 14 (5) (2010) 754−769, doi.org/10.1111/j.1530-9290.2010.00269.x.

[36] S. George, D. McKay, The Environmental Impact of Music: Digital Records, CDs Analysed. The Conversation − Environment + Energy, Posting Dated 10 January 2019, 2019. theconversation.com/the-environmental-impact-of-music-digital-records-cds-analysed-108942.

[37] MusicTank, The Dark Side of the Tune, Posting Dated 13 September 2016, 3 pages. www.musictank.co.uk/wp-content/uploads/2016/03/EnergyReportPR.pdf.

[38] M.P. Mills, The Cloud Begins With Coal − Big Data, Big Networks, Big Infrastructure, and Big Power − an Overview of the Electricity Used by the Global Digital System, Report Sponsored by the National Mining Association and the American Coalition for Clean Coal, August 2013, 2013, 45 pages, www.tech-pundit.com/wp-content/uploads/2013/07/Cloud_Begins_With_Coal.pdf.

[39] L. Desjardin, Video Killed the Energy Star: Why 5G Must Use Less Power, EDN, published by Aspencore, Cambridge Massachusetts, posted on 20 February 2018, www.edn.com/video-killed-the-energy-star-why-5g-must-use-less-power/.

[40] Netflix, Environmental Social Governance, 2019, 9 pages, s22.q4cdn.com/959853165/files/doc_downloads/2020/02/0220_Netflix_EnvironmentalSocialGovernanceReport_FINAL.pdf.

[41] K. Mayers, J. Koomey, R. Hall, M. Bauer, C. France, A. Webb, The carbon footprint of games distribution, J. Ind. Ecol. 19 (3) (2014) 402–415, doi.org/10.1111/jiec.12181.

[42] D. Quack, M. Oley, Environmental advantages of video conferencing systems — Results from a simplified LCA, in: Proc. EnviroInfo, 16th Conf. Environmental Communication in the Information Society, Vienna IGU/ISEP, 2002, pp. 446–450. enviroinfo.eu/sites/default/files/pdfs/vol106/0446.pdf.

[43] C. Weber, C. Hendrickson, P. Jaramillo, S. Matthews, A. Nagengast, R. Nealer, Life Cycle Comparison of Traditional Retail and E-Commerce Logistics for Electronic Products: A Case Study of buy.Com. ISSST '09, IEEE Int. Symposium on Sustainable Systems and Technology, 2009, pp. 1–6, doi.org/10.1109/ISSST.2009.5156681.

[44] R. Mangiaracina, G. Marchet, S. Perotti, A. Tumino, A review of the environmental implications of B2C e-commerce: A logistics perspective, Int. J. Phys. Distrib. Logist. Manag. 45 (6) (2015) 565–591, doi.org/10.1108/IJPDLM-06-2014-0133.

[45] R.J. Caudill, Y. Luo, P. Wirojanagud, M. Zhou, A lifecycle environmental study of the impact of e-commerce on electronic products, in: Proc. 2000 IEEE Int. Symposium on Electronics and the Environment (Cat. No.00CH37082), San Francisco, 2000, pp. 298–303, doi.org/10.1109/ISEE.2000.857665.

[46] H.S. Matthews, C.T. Hendrickson, D.L. Soh, Environmental and economic effects of e-commerce: A case of book publishing and retail logistics, J. Transp. Res. Board Rec. 1763 (2001) 6–12, doi.org/10.3141/1763-02.

[47] H.S. Matthews, E. Williams, T. Tagami, C.T. Hendrickson, Energy implications of online book retailing in the United States and Japan, Environ. Impact Assess. Rev. 22 (5) (2002) 493–507, doi.org/10.1016/S0195-9255(02)00024-0.

[48] J.W.J. Weltevreden, O. Rotem-Mindali, Mobility effects of b2c and c2c e-commerce in The Netherlands: a quantitative assessment, J. Transp. Geogr. 17 (2) (2009) 83–92, doi.org/10.1016/j.jtrangeo.2008.11.005.

[49] A. Halldórsson, J.B. Edwards, A.C. McKinnon, S.L. Cullinane, Comparative analysis of the carbon footprints of conventional and online retailing, Int. J. Phys. Distrib. Logist. Manag. 40 (1–2) (2010) 103–123, doi.org/10.1108/09600031011018055.

[50] M. Pärssinen, M. Kotila, R. Cuevas, A. Phansalkar, J. Manner, Environmental impact assessment of online advertising, Envl. Impact Assess. Rev. 73 (2018) 177–200, doi.org/10.1016/j.eiar.2018.08.004.

[51] G.G. Parker, M.W. Van Alstyne, S.P. Choudary, Platform Revolution: How Networked Markets Are Transforming the Economy - and How to Make Them Work for You, W. W. Norton & Co., 2016, 336 pages.

[52] L. Bliss, Ride-hailing Isn't Really green, Bloomberg BusinessWeek, 2020, 25 February 2020, www.bloomberg.com/news/articles/2020-02-25/the-other-toll-of-uber-and-lyft-rides-pollution.

[53] J. Martin, Five Things You Should Know about Lyft and Uber's Climate Impacts (And what You Can Do), Union of Concerned Scientists, 2020 posting dated 25 February 2020, blog. ucsusa.org/jeremy-martin/five-things-you-should-know-lyft-and-ubers-climate-impacts; For Exact Numbers, See: Union of Concerned Scientists — Transportation - Ride-Hailing Is a Problem for the Climate. Here's Why. Feature dated February 2020. www.ucsusa.org/resources/ride-hailing-problem-climate.

[54]   eBay Inc, Impact Report, 2019. April 2020, 43 pages, static.ebayinc.com/assets/Uploads/ Documents/eBay-Impact-2019-Report.pdf.

[55]   Etsy News, Etsy Becomes the First Global eCommerce Company to Completely Offset Carbon Emissions from Shipping, by Josh Silverman, February 26, 2019. blog.etsy.com/ news/2019/on-etsy-every-purchase-makes-a-positive-impact/.

Chapter 9

Humans and their needs

9.1 Population

The size of the human population on the earth has increased and is expected to increase every decade as tabulated in the following.

World population over time		
Year	**World population**	**Source**
1700 (estimated)	600,000,000	[1]
1800 (estimated)	990,000,000	[1]
1900 (estimated)	1,650,000,000	[1]
2000	6,143,493,806	[2]
2010	6,956,823,588	[2]
2020	7,794,799,000	[2]
2030 (projected)	8,548,487,000	[2]
2040 (projected)	9,198,847,000	[2]
2050 (projected)	9,735,034,000	[2]
2060 (projected)	10,151,470,000	[2]
Sources: [1,2].		

For the population of individual countries, a reliable and up-to-date source is the CIA World Factbook [3].

The population density averaged over the land surface of the earth (excluding Antarctica) was 57 people/km^2 (38 people per square mile) in 2019 [4]. It is very unevenly distributed as revealed by the following table.

Data, Statistics, and Useful Numbers for Environmental Sustainability.
https://doi.org/10.1016/B978-0-12-822958-3.00007-8

Population density across the world		
Continent	**People/km^2**	**People/square mile**
Asia	99.6	257.8
Europe	72.5	187.7
Africa	40.1	103.7
South America	23.7	61.3
North America	23.4	60.7
Australia	3.0	7.8
Source: [4].		

9.2 Types of footprints

To meet their personal needs, people affect their environment in multiple ways, particularly by energy consumption and carbon emissions. According to an analysis performed for the United Nations Environment Programme, the annual carbon footprint of households in developed countries ranges widely from about 3.8 (in frugal Switzerland) to 22.4 (in the consumptive United States) metric tons of CO_{2eq} per capita ([5] Fig. 4.3). The relative contributions to the carbon footprint by type of consumptive activity are tabulated in the following.

Human carbon footprint by need	
Personal need	**Fraction of carbon footprint**
Shelter	26%
Food	27%
Mobility	20%
Services	9%
Product manufacturing	7%
Commerce	7%
Clothing	4%
Source: [5] Table 4.2.	

9.3 Shelter

As a rule, the higher the local population density, the smaller the space in which people live. There is also some variation across countries that may reflect cultural values and availability of land. The following table compares the average residential floor space in various places.

Average residential space in various places		
	Floor space per person	
Country/City	in m²	in ft²
Australia	89	960
Canada	72	779
China	20	215
Denmark	65	702
France	43	464
Germany	54.5	587
Greece	45	484
Hong Kong	15	161
Italy	31	335
Japan	35	379
New York City	49.3	531
Russia	22	237
Spain	34.7	373
Sweden	39.5	425
United Kingdom	33	356
United States	73.3	789
Sources: [6–8].		

In the countryside, people live in larger spaces but also have larger households, with still more space per person, at least in the United States as shown in the following.

Residential floor space by location and household size			
Location	Median floor space	People per household	Median space per person
Urban	1,931 ft²	2.53	762 ft²
Rural	2,320 ft²	2.60	891 ft²
USA overall	2,008 ft²	2.54	789 ft²
Source: [8], 2015 data.			

The average single-family house built in 2000 in the United States required 19 tons of concrete, 1,153 ft³ of lumber, and 3,061 ft² of insulation [9].

9.4 Food

According to food labels, a human adult requires a daily food intake of 2,000 food calories. This amount is equal to 2,000 kcal $= 2 \times 10^6$ calories $= 7,940$ BTUs $= 8,370$ kJ. On a per-time basis, this is equivalent to a power consumption of about 97 W/person. This number, however, is very approximate as caloric needs vary with age, gender, and level of activity, as the following table indicates.

Food needs in kcal per person per day				
	Male		Female	
Age group	Moderately active	Active	Moderately active	Active
Child	1,750	2,000	1,600	1,800
Teenager	2,680	3,100	2,000	2,400
Adult	2,600	2,900	2,000	2,200
Senior	2,300	2,600	1,800	2,000

Source: [10].

There are 3,500 kcal in 1 lb (0.454 kg) of fat, and one, therefore, needs to burn 3,500 kcal to lose 1 lb [11].

In the United States, the energy consumed in food production, packaging, delivery, and preparation is divided as follows.

How energy is spent for food production	
Stage in food chain	Fraction of energy
Agricultural production	14%
Processing industry	19%
Packaging	6%
Transportation	4%
Wholesale and retail	16%
Food services	13%
Household storage and preparation	28%
Total	100%

Source: [12].

For the average American diet, it has been estimated that it takes 14.2 quad (=15.0 EJ) of energy to produce 1.75 quad (=1.85 EJ) of food, of which only 1.07 quad (=1.13 EJ) is consumed, the rest (39%) being wasted [12]. The ratio 14.2:1.75 indicates that it takes 8.1 times as much energy to produce, process, package, transport, refrigerate, and prepare food as there is energy in the food itself. The ratio 14.2:1.07 may be interpreted as an efficiency ratio: It takes 13 calories of energy for every 1 calorie of intake. Since the human metabolism is only 20% efficient at best [13], it takes 66 J of energy to produce the food that will enable a person to perform 1 J of mechanical work.

The following table lists the amount of energy needed to produce the most common foods.

Production energy and energy content of various foods				
	Energy to produce		Energy content		Efficiency:content/ production
Food	kWh/kg	kWh/lb	kcal/kg	kcal/lb	
Apples	3.68	1.67	476	216	15%
Beef	69.4	31.5	2,593	1,176	4.3%
Cheese	14.9	6.75	4,021	1,824	31%
Chicken	9.7	4.4	1,263	573	15%
Corn	0.95	0.43	860	390	102%
Eggs	8.8	4.0	1,433	650	19%
Milk	1.65	0.75	642	291	45%
Pork	27.8	12.6	1,058	480	8.5%

Source:[14].

A longer list of grocery items with their energy and water footprint follows.

Energy and water footprints of common grocery items			
Group	Item	Energy (MJ/kg)	Water footprint impact indicator[1]
Dairy	Butter	32.00	330,000
	Cheese	51.80	512,000
	Cream	–	58,500
	Eggs	27.20	350,000
	Ice cream	15.00	–
	Milk—fresh	5.10	1,840,000
	Milk—concentrate/ powder	62.50	62,300
	Yogurt	13.65	206,000

Continued

Meats and fish	Beef	50.71	1,260,000
	Fish and seafood—canned	19.50	—
	Fish and seafood—chilled	54.50	—
	Fish and seafood—frozen	94.90	—
	Lamb	67.00	412,000
	Pork	35.10	632,000
	Poultry	40.35	646,000
Vegetables	Beans (fresh)	90.30	—
	Carrots	2.80	—
	Cucumbers	42.00	—
	Lettuce	6.00	—
	Mushrooms	47.63	—
	Onions	2.90	162,000
	Potatoes	2.10	142,000
	Rice	14.93	383,000
	Tomatoes	42.39	—
	Vegetables—canned	17.35	127,000
	Vegetables—frozen	19.50	65,000
Fruits	Apples	4.74	247,000
	Bananas	5.40	660,000
	Canned fruits	13.00	—
	Grapes	8.75	131,000
	Oranges	—	131,000
	Strawberries	12.70	—
	Other fresh fruit	23.00	69,800
Beverages	Beer—ale	18.75	—
	Beer—lager	3.50	115,000
	Cider	—	561,000
	Coffee	74.50	1,850,000
	Juices	10.20	802,000
	Juice concentrates	—	3,920,000

	Spirits	82.50	136,000
	Tea	65.45	559,000
	Water (bottled)	2.92	—
	Wines	14.00	449,000
	Wines—sparkling	36.38	—
	Bread	12.10	1,010,000
	Cakes and pastries	16.00	502,000
	Candy (sugar confections)	34.00	415,000
	Cereals (breakfast type)	15.50	144,000
	Chocolate	43.00	11,200,000
	Crackers	15.50	—
	Cookies (biscuits)	23.00	—
	Margarine	23.20	212,000
Other food items	Morning pastries	18.50	—
	Nuts and seeds	—	114,000
	Pasta (dried)	14.50	131,000
	Pizza (frozen)	22.54	—
	Potato chips	37.00	—
	Sauces	15.96	—
	Seasonings	—	381,000
	Snacks (processed)	14.59	—
	Soup (canned)	18.46	—
	Cat food	15.45	—
	Dog food	15.45	—
	Bath and shower products	13.65	—
	Cleaning products	16.15	—
	Detergent—liquid	31.05	—
Nonfood items	Detergent—powder	30.42	—
	Diapers (nappies)	82.50	—
	Toilet care	66.85	—
	Toilet paper	20.75	—

[1]The water footprint impact indicator is defined as a scarcity-weighted number of liters per kg of product [15].
Source: [15] Tables 4.1 (energy) and 5.1 (water).

A carton of 12 eggs has a carbon footprint of 3.6 kg (7.9 lb) of CO_{2eq}. The carbon footprint of 1 kg of tomatoes varies greatly depending on whether they are grown in season on a local field or in a greenhouse in a faraway place, from 0.4 kg (0.9 lb) CO_{2eq} for the former to 50 kg (110 lb) CO_{2eq} for the latter; an average value of 9.1 kg (20 lb) CO_{2eq} is reasonable ([16] pages 98−99). Additional food carbon footprints can be found in Section 4.3.7.

The conventional production of a $^1/_4$ -lb hamburger requires 6.2 kg (=13.6 lb) of grain and forage, 55.1 L (=14.5 gallons) of water, 6.19 m^2 (=66.6 ft^2) of land, and 995 kJ (=943 BTUs) of fossil fuel, which is enough energy to power a typical microwave for 16 minutes. It is also responsible for 1.81 kg CO_{2eq} of greenhouse gas emissions. All these numbers are higher for natural and grass-fed cattle rearing ([17] Table 2, prorated from 1 billion kg of beef to a $^1/_4$ -lb hamburger).

A life-cycle analysis of bread making in a large city of Indonesia [18] reveals that the production of 1 kg of bread from wheat and eggs on the farm to the packaged product at the exit of the bakery necessitates the consumption of 241 kJ and is responsible for the emission of 2.56 kg CO_{2eq} (74% of which is due to agriculture). A Danish study [19] estimates the energy demand to be significantly higher, 6.2 MJ/kg for an industrial bakery and 8.7 MJ/kg for home baking, and the carbon footprint significantly lower, 520 and 680 g CO_{2eq}/kg, respectively. This study also estimates that it takes 1.1 m^2 of farmland to produce 1 kg of bread with conventional farming practices, and 1.7 m^2 of farmland using organic farming practices. Taking mid-values between these studies, we estimate that it takes of 1,460 kJ and 700 g CO_{2eq} for the typical 1-lb loaf of bread. This carbon footprint is equivalent to driving 1 automobile for 1.4 hours.

The American diet amounts to 25.2 MJ of nonrenewable energy demand with a 4.7 kgCO_{2eq} carbon footprint per person per day [12].

9.5 Clothing

Clothing is not without some significant carbon footprint, counting for about 4% of the personal footprint of people ([5] Table 4.2). In the United Kingdom, which may be representative of most highly developed countries, textiles and clothing account for 2% of the country's total carbon footprint ([16] Fig. 11.2, page 163).

These footprints are in part due to the manufacturing and weaving of the textile, dyeing it, and then sewing the garment. See Section 12.12 for numbers pertaining to the textile industry.

Carbon footprint of various types of clothing			
Article of clothing	CO_{2eq} per item over lifetime		Average number of uses/washes	Source
Basic T-shirt	2.3 kg	5.1 lb	22/11	[20] Table 4 and Fig. 6
Dress	16.9 kg	37.2 lb	10/3.33	[20] Table 4 and Fig. 6
Polyester blouse[1]	3.7 kg	8.1 lb	40/20	[21] Table 1
Long-sleeve cotton shirt	10.75 kg	23.7 lb	−/50	[22] Fig. 7.11
Pair of cotton jeans	10.7 kg	23.5 lb	200/20	[20] Table 4 and Fig. 6
Pair of shoes—average	11.5 kg	25.3 lb	−	[16] p. 105
Pair of shoes—leather	15 kg	33 lb	−	[16] p. 105
Jacket	17.2 kg	37.8 lb	100/1	[20] Table 4 and Fig. 6
Fleece jacket with hood	13.4 kg	29.6 lb	−	[22] Fig. 7.16
Children's jacket	13.7 kg	30.3 lb	−	[22] Fig. 7.16
Hospital uniform	9.9 kg	21.8 lb	75/75	[20] Table 4 and Fig. 6

[1]*Carbon footprint inferred from energy demand at 5.74×10^{-5} kg CO_{2eq} per BTU.*
Sources: [16,20−22] as indicated in right column.

Laundering of washable items is a major addition to the footprint because of the energy needs for warm water and electric drying [21]. The carbon footprint is generally more massive than the garment itself. For example, the modest 170-g T-shirt is responsible for an emission of 2.3 kg of CO_{2eq}, 13 times its own mass.

The following table provides the energy requirements for a polyester blouse, assuming 40 wearings and 20 washings over its lifetime. We note that laundering amounts to 82% of the total energy expenditure. This can be reduced by 93% by cold wash and air drying ([21] Fig. 8.3). The production of the blouse can be reduced by switching to a less impacting fiber (see series of tables in Section 12.12).

Energy contributions to the life-cycle of a polyester blouse			
	Energy		
Life stage	kJ/blouse	BTUs/blouse	fraction	
Resin manufacturing	5,237	4,964	7.72%	
Fiber manufacturing	848	804	1.25%	
Dye	215	204	0.32%	
Fabric manufacturing	4,777	4,528	7.04%	18.2%
Apparel—production	34	32	0.05%	
Apparel—packaging	831	788	1.23%	
Apparel—transportation	430	408	0.63%	
Detergent manufacturing	1,713	1,624	2.43%	81.7%
Laundry	53,723	50,920	79.2%	
Disposal	34	32	0.05%	0.05%
Total	67,844	64,304	100%	100%

Note: Manufacturing stages include material procurement, intermediate packaging, and transportation.
Source: [21] Table 1, with numbers converted from 10^6 wearings to a single blouse.

A typical household laundry load is 3.5 kg (=7.7 lb) per wash cycle ([23] Table 1). In the United States, a household does about 289 washing cycles per year, and each washing cycle consumes 0.43 kWh of electricity and 144 L (=38.0 gal) of water. Across Europe and most of the developed world, the typical numbers are 177 cycles/year, 0.95 kWh/cycle, and 60 L/cycle ([23] Table 2).

The US Environmental Agency gives its Energy Star label to commercial washing machines that have a Minimum Energy Factor (MEF) of at least 2.0, defined as [24].

$$\text{MEF} = \frac{\text{Volume capacity of drum (in ft}^3)}{\text{Electricity use + energy for hot water (both in kWh/cycle)}}$$

$$\geq 2 \frac{\text{ft}^3}{\text{kWh/cycle}}.$$

9.6 Household activities

A compilation of carbon footprints of the average household in the United Kingdom [25] revealed an estimated total of 26 metric tons of CO_{2eq} for the year 2004. The breakdown is tabulated in the following.

Carbon footprint of a UK household by item and activities	
Item or Activity	Fraction
Recreation and leisure	27%
Food and Catering	24%
Space heating	13%
Household items	11%
Health and hygiene	9%
Clothing and footwear	8%
Commuting	5%
Education	2%
Communications	1%
Total	26 metric tons CO_{2eq}/year
Source: [25].	

9.7 Human energy

By virtue of its metabolism, the human body produces heat continuously, the rate of which increases with the level of activity. The following table provides the approximate energy expenditure for a series of common activities.

Energy expenditure of humans according to activity			
	Energy expenditure		
Activity	Watts	BTUs/hour	kcal/minute
Sleeping	100	340	1.43
Sitting or lying awake	84	285	1.2
Standing	90	310	1.3
Walking—moderate	244	830	3.5
Walking—brisk	860	2,930	12.3
Bicycling—moderate	280	950	4.0
Bicycling—fast	750	2,550	10.7
Light work	200	680	2.9
Jogging	800	2,720	11.5
Running	1,000	3,400	14.3
Swimming—moderate	380	1,300	5.5
Sources: [26] page 67, [27,28].			

An easy number to remember: A runner who weighs 78 kg (=170 lb) burns 78 kcal/km [28].

The following table is similar except that the energy expenditure refers to the occupation of the person while at work.

Energy expenditure of humans according to occupation			
	Energy expenditure		
Occupation	Watts	BTUs/hour	kcal/minute
Administrative work	154	524	2.20
Blacksmith	262	894	3.75
Carpenter	257	878	3.69
Crane operator	190	647	2.72
Gardener	275	940	3.95
Glazier	194	662	2.78
Machinery operator	176	601	2.52
Mason	244	832	3.50
Miller	199	678	2.85
Painter	208	709	2.98
Turner	181	616	2.59
Welder	181	616	2.59

Source: [29] page 247.

Sources

[1] Our World in Data, World Population, by Max Roser, Hannah Ritchie and Esteban Ortiz-Ospina, Substantially Revised in May 2019. ourworldindata.org/world-population-growth.

[2] United Nations, Department of Economic and Social Affairs — Population Dynamics — World Population Prospects (2019). Past and current: population.un.org/wpp/Download/Standard/Population/ Projections: population.un.org/wpp/Download/Probabilistic/Population/.

[3] U.S. Central Intelligence Agency (CIA), Libray — Publications — the World Factbook. www.cia.gov/library/publications/the-world-factbook/.

[4] M. Rosenberg, Population Density Information and Statistics, 2019 updated 9 May 2019, www.thoughtco.com/population-density-overview-1435467.

[5] U. N. Environment Programme (UNEP), International Panel for Sustainable Resource Management — Assessing the Environmental Impacts of Consumption and Production: Priority Products and Materials, 2010, 112 pages. ISBN: 978-92-807-3084-5, www.unep.fr/shared/publications/pdf/DTIx1262xPA-PriorityProductsAndMaterials_Report.pdf.

[6] L. Wilson, How Big Is a House? Average House Size by Country. ShrinkThatFootprint, Posted Circa November 2014, 2014. shrinkthatfootprint.com/how-big-is-a-house.

[7] ThatcherClay, Residential density in NYC, urban_calc, posted 28 February 2017. urbancalc. com/post/NYC-Residential-Density/.

[8] U.S. Energy Information Administration, Residential Energy Consumption Survey (RECS) — Table HC10.9 Average Square Footage of U.S. Homes, 2015 (revised May 2018), www.eia. gov/consumption/residential/data/2015/hc/php/hc10.9.php.

[9] University of Michigan, Cnter for Sustainable Systems — Residential Buildings Factsheet, Pub. No. CSS01-08, 2019. css.umich.edu/factsheets/residential-buildings-factsheet.

[10] U.S. Department of Health and Human Services, Dietary Guidelines 2015-2020 — Appendix 2. Estimated Calorie Needs Per Day, by Age, Sex, and Physical Activity Level. health.gov/our-work/food-nutrition/2015-2020-dietary-guidelines/guidelines/appendix-2/.

[11] Mayo Clinic — Healthy Lifestyle, Counting Calories: Get Back to Weight-Loss Basics, posted 2 July 2020. www.mayoclinic.org/healthy-lifestyle/weight-loss/in-depth/calories/art-20048065.

[12] University of Michigan — Center for Sustainable Systems — U.S, Food System Factsheet, Pub. No. CSS01-06, 2019. css.umich.edu/factsheets/us-food-system-factsheet.

[13] Engineering-abc.com, Energy Consumption during Cycling. www.tribology-abc.com/calculators/cycling.htm.

[14] The Oil Drum, List of Foods by Environmental Impact and Energy Efficiency, Posted by Gail the Actuary on 2 March 2010. www.theoildrum.com/node/6252.

[15] Product Sustainability Forum — Environmental Performance of Products — an Initial Assessment of the Environmental Impact of Grocery Products, March 2013, 135 pages, www.wrap.org.uk/sites/files/wrap/An%20initial%20assessment%20of%20the%20environmental%20impact%20of%20grocery%20products%20final_0.pdf.

[16] M. Berners-Lee, How Bad Are Bananas? The Carbon Footprint of Everything, Greystone Books, D&M Publishers, 2011, 232 pages.

[17] J.L. Capper, Is the grass always greener? Comparing the environmental impact of conventional, natural and grass-fed beef production systems, Animals 2 (2012) 127—143, doi.org/10.3390/ani2020127.

[18] N.H. Laurence, A. Christiani, Case study of life cycle assessment in bread production, in: Proceedings of 2nd Int. Conf. On Eco Engineering Development (ICEED 2018), IOP Conf. Ser.: Earth and Environmental Sci, vol. 195, 2018, p. 012043, doi.org/10.1088/1755-1315/195/1/012043.

[19] G.A. Reinhardt, J. Braschkat, A. Patyk, M. Quirin, Life cycle analysis of bread production — a comparison of eight different options, in: Lecture Slides Presented at the 4th International Conference on Life Cycle Assessment in the Agri-Food Sector, Horsens, Denmark, 6—8 October 2003, 2003, 28 slides, www.lcafood.dk/lca_conf/contrib/g_reinhardt.pdf.

[20] S. Roos, G. Sandin, B. Zamani, G. Peters, Environmental Assessment of Swedish Fashion Consumption. Five Garments — Sustainable Futures. Report Prepared for Mistra Future Fashion, 2015, 15 June 2015, 142 pages. mistrafuturefashion.com/wp-content/uploads/2015/06/Environmental-assessment-of-Swedish-fashion-consumption-LCA.pdf.

[21] G.G. Smith, R.H. Barker, Life cycle analysis of a polyester garment, Res. Conserv. Recycling 14 (3—4) (1995) 233—249, doi.org/10.1016/0921-3449(95)00019-F.

[22] S. Rana, S. Pichandi, S. Karunamoorthy, A. Bhattacharyya, S. Parveen, R. Fangueiro, Carbon footprint of textile and clothing products, Chap. 7 in S.S. Muthu (Ed.), Handbook of Sustainable Apparel Production, Taylor & Francis, 2015, pp141—165, doi.org/10.1201/b18428-10.

[23] C. Pakula, R. Stamminger, Electricity and water consumption for laundry washing by washing machine worldwide, Energy Eff. 3 (2010) 365–382, doi.org/10.1007/S12053-009-9072-8.

[24] Energy- and Water-Saving Fact Sheet, Commercial Washing Machines – Energy and Water Energy Star Benchmarks, January 2012. Commercial_Washing_Machines_Fact_Sheet-2.pdf.

[25] A. Druckman, T. Jackson, An Exploration into the Carbon Footprint of UK Households, RESOLVE Working Paper 02-10, Centre for Environmental Strategy (D3), University of Surrey, 2010, 35 pages. resolve.sustainablelifestyles.ac.uk/sites/default/files/RESOLVE_WP_02-10.pdf.

[26] N. Lechner, Heating, Cooling, Lighting, fourth ed., Wiley, 2015, 702 pages.

[27] F.J. Amaro-Gahete, G. Sanchez-Delgado, J.M.A. Alcantara, B. Martinez-Tellez, F.M. Acosta, E. Merchan-Ramirez, M. Löf, I. Labayen, J.R. Ruiz, Energy expenditure differences across lying, sitting, and standing positions in young healthy adults, PLoS One 14 (6) (2019) e0217029, doi.org/10.1371/journal.pone.0217029.

[28] B. MacKenzie, Energy expenditure, BrainMAC Sports Coach, (2002). www.brianmac.co.uk/energyexp.htm#ref. Making Reference to a Book, of Which a Later Edition Is: W.D. McArdle, F.I. Katch, V.L. Katch, Essentials of Exercise Physiology, fourth ed., Lippincott Williams & Wilkins, (2010) 790 pages.

[29] A. Martínez-Rocamora, J. Solís-Guzmán, M. Marrero, Carbon footprint of utility consumption and cleaning tasks in buildings, Chap. 9 in S.S. Muthu (Ed.), Environmental Carbon Footprints – Industrial Case Studies, Butterworth-Heinemann, Elsevier, 2018, pp. 229–258, doi.org/10.1016/B978-0-12-812849-7.00009-X.

Chapter 10

Risks

Risk is the probability that an adverse effect from some activity or exposure will occur over a given period of time. This period of time is usually taken as the duration of an event, one year, or a lifetime. To switch from one time unit to another, one uses the life expectancy of a person, which varies with country, gender, and many other factors. In the United States, the overall expectation of life at birth was 78.7 years in 2018 [1]. While exposure is usually for a lifetime, specific activities may be pursued only during adult years, in which case childhood years must be discounted from the lifetime.

10.1 Units

In the environmental context, a risk is usually a probability per time, such as the time spent in a risky activity or over a lifetime. For specific activities, it may be expressed as a probability per amount of the activity, such as per mile driven in an automobile or per volume of alcohol consumed.

10.2 Lifetime risks

We shall all die one day. One question concerns the cause of our future death. The chances that the average American will die of one cause or another can be cast as odds, as follows.

Odds of dying in the United States in decreasing level of risk	
Cause of death	**Lifetime odd[1]**
Heart disease	1 in 4.3
Cancer (malignant neoplasm)	1 in 4.7
Accident	1 in 17.0
Motor vehicle traffic accident	1 in 77.7
Poisoning by drug overdose	1 in 42.8
Opioid overdose	1 in 98
Unintentional fall	1 in 111

Continued

Data, Statistics, and Useful Numbers for Environmental Sustainability.
https://doi.org/10.1016/B978-0-12-822958-3.00013-3

As a pedestrian	1 in 541
As a bicyclist	1 in 4,060
Airplane crash (if you are a passenger[2])	1 in 5.4 million
Chronic lower respiratory disease	1 in 17.8
Stroke	1 in 19.2
Alzheimer's disease	1 in 23.3
Diabetes	1 in 33.4
Influenza or pneumonia	1 in 48.0
Kidney disease	1 in 55.3
Suicide	1 in 58.7
Chronic liver disease/Cirrhosis	1 in 67.4
Septicemia	1 in 68.8
Hypertension (excluding heart disease)	1 in 79.7
Parkinson's disease	1 in 88.0
Murder (homicide)	1 in 175
Gun assault	1 in 298
AIDS and HIV	1 in 498
Drowning	1 in 1,120
Fire or smoke	1 in 1,400
Choking on food	1 in 2,600
Childbirth (if you are a mother[2]) (17.4 maternal deaths in 100,000 live births)	1 in 5,750
Accidental gun discharge	1 in 9,000
Killed by an animal (horse being likely)	1 in 34,000
Hornet, wasp, and bee stings	1 in 54,000
Attack by a dog	1 in 119,000
Strike of lightning	1 in 181,000
Any cause	1 in 1, obviously

[1]Calculated as total number of death in the year divided by the number of deaths from that particular cause.
[2]Parenthetical conditions here and elsewhere indicate that the risk is calculated for that particular population rather than for the general population.
Sources: [1] Data table for Fig. 2, [2] Table B, [3–10] whichever has the most recent data.

The risk of dying in an airplane crash being 1 in 5.4 million implies that one would expect to be killed in an airplane crash if one were flying once every day for 15,000 years [10].

Many of life's happenstances are non-fatal, including sicknesses, aggression, and exposure to environmental hazards. Here are some annual and lifetime odds.

Risks of non-fatal situations		
Event	Odd in 1 year	Odd in 78 years
Hospitalization due to seasonal flu	1 in 1,180	1 in 7.8
Hospitalization due to water-borne pathogen	1 in 680	1 in 8.7
Hospitalization caused by air pollution in Boston (Suffolk Co.)	1 in 790	1 in 10
Victim of aggravated assault	1 in 15	1 in 15

Sources: [11−14].

The following activities carry the respective risks.

Risky activities, in decreasing level of risk		
Activity	Annual risk (deaths per year per 100,000 persons at risk)	Source
Smoking (all causes)	978	[15]
Riding a motorcycle	59.3	[16]
Farming	20.4	[17]
Coal mining	15	[18]
Riding in a motor vehicle	11.2	[19]
Drinking chlorinated water	0.8	[20] p. 120
Eating a 3 oz. (85 g) charcoaled steak every day	0.5	[20] p. 120
Reference: One-in-a-million lifetime risk	0.00143	$=100,000/10^6 \times$ 1/70 years

Sources: As indicated in the last column.

The preceding table ends with the risk of one-in-a-million lifetime risk. This is the level that the Health and Safety Executive in the United Kingdom deems to be an "acceptable risk" at which no further improvements in safety need to be made [21]. This is also the lower level of risk among two that the US Environmental Protection Agency deems acceptable for carcinogens in drinking water [21]. Permitting one person to die among a million in a lifetime of 70 years (or 1 person dying every year among 70 million) may appear immoral to some, but a limit of the tolerable had to be set in a world where nothing is ultimately risk-free. The following section compares activities that all lead to this level of risk, and it shows that the 1-in-a-million in a lifetime is actually a very low risk, one that most of us dare exceed on a frequent basis.

10.3 Activities of equal risks

The following activities all carry the same risk of 1-in-a-million during lifetime.

Activities with 1-in-a-million lifetime risk		
Activity	Nature of the risk	1 in a million chance of dying in 70 years
Smoking	Cancer, heart disease	0.57 cigarettes
Living with a cigarette smoker	Cancer, heart disease from 2nd-hand smoke	The smoke of 6.1 cigarettes
Drinking diet soda	Cancer caused by saccharin	30 12-oz cans (10.6 L)
Drinking alcohol	Cirrhosis of liver	0.55 L of wine 0.19 L of whiskey
Rock climbing	Fatal fall	1.5 min on the rock face
Downhill skiing	Fatal injury	46 min on the slope
Living in a large, polluted city	Air pollution	8.5 hours
Living at high elevation (*Ex.* Denver)	Cancer from cosmic radiation	2 months
Riding a bicycle	Fatal accident	130 km or 5.2 hours
	Nonfatal injury	0.54 km or 1.1 minutes
Riding a motorcycle	Fatal accident	6.3 km
	Nonfatal injury	0.36 km
Traveling by car	Fatal accident	133 miles = 215 km
Flying with a commercial aircraft	Airplane crash after takeoff	5.4 flights
	Cancer from cosmic radiation	6,000 miles = 9,700 km
Canoeing	Drowning	6 minutes
Eating moldy peanut butter	Liver cancer from aflatoxin	40 tablespoons = 640 g

Eating charcoal-broiled beef	Cancer from benzopyrene	100 steaks
Living near a nuclear power reactor	Accidental radioactivity release	50 years

Sources: [22] or similar calculation performed on more recent statistics, [23] page 43.

When someone engages in multiple activities, the independent risks simply add up. For example, a person's vacation that entails 2,000 miles of roundtrip flying, 400 miles of traveling by car, rock climbing for 3 hours, canoeing for 2 hours, drinking 1 L of wine, and eating 2 charcoal-broiled steaks carries a compounded risk of (2 flights)/(5.4 flights) + (2,000 mi)/(6,000 mi) + (400 mi)/(133) + (180 min)/(1.5 min) + (120 min)/(6 min) + (1 L)/(0.55 L) + (2 steaks)/(100 steaks) = 146 chances in a million of dying prematurely.

10.4 Environmental risk assessment

The determination of an environmental risk begins with the estimation of the level of exposure to a potentially harmful substance, called the *Chronic Daily Intake* (*CDI*), expressed in mg/(kg.day). It is calculated from a series of factors:

$$CDI = C \frac{CR \times EF \times ED}{BW \times AT},$$

In this formula, C is the concentration of the substance (in mg/L in water or mg/m^3 in air), CR is the contact rate (in L of water drunk per day or m^3 of air inhaled per day), EF is the exposure frequency (in days per year), ED is the exposure duration (in years), BW is the body weight (in kg), and AT is the time period over which the exposure is averaged (in days). If specific values are not known for some of the factors, default values for a maximally exposed individual are used as tabulated in the following.

	Default values for a maximally exposed individual		
Factor	Adult at home	Child	Worker
CR	2 L water/day 20 m^3 air/day	1 L water/day 12 m^3 air/day	1 L water/workday 10 m^3 air/workday
EF	350 days/year		245 workdays/year

Continued

ED	Actual event duration or 30 years if chronic		Actual event duration or 21 years if chronic
BW	70 kg	15 kg	70 kg
AT non-carcinogenic substance	Same as ED		
AT carcinogenic substance	365 days/year × 70-yr lifetime		
Source: [24].			

10.4.1 Noncancer risk

For noncancer risks, there is an exposure threshold below which there is no harmful effect to humans no matter how repeated the exposure may be. Because of unavoidable uncertainties, a value somewhat lower than this threshold is adopted as the *No Observed Adverse Effect Level*. For added safety, one or several divisions by 10 are then applied to obtain a tolerable level [25].

For oral ingestion, this tolerable level takes the form of a *Reference Dose* (*RfD*) expressed in milligrams of the substance per kilogram of body weight per day. For inhalation, the tolerable level is stated as a *Reference Concentration* (*RfC*) expressed in milligrams of the chemical per m^3 of air. The following table lists the *RfD* and *RfC* for a sample of harmful substances. Note that for carcinogens like benzene and 1,1,1-trichloroethane the values listed here are for the evaluation of their risk of poisoning. For their cancer risk, which is a separate risk, see the next subsection.

		Ingested orally: reference dose	Inhaled: reference
Substance	Chemical formula	(RfD) mg/ (kg.day)	concentration (RfC) mg/m^3
Acetone	C_3H_6O	0.9	−
Acrylic acid	$C_3H_4O_2$	0.5	0.001
Ammonia	NH_3	−	0.5
Atrazine	$C_8H_{14}ClN_5$	0.035	−
Benzene	C_6H_6	1.2	8.2
Bisphenol A	$C_{15}H_{16}O_2$	0.05	−
Cadmium	Cd	5×10^{-4}	−

Reference doses and concentrations for noncarcinogenic risk assessment

Carbon tetrachloride	CCl_4	4×10^{-3}	0.1
Chlorine	Cl	0.1	—
Chlorine dioxide	ClO_2	0.03	2×10^{-4}
Chloroform	$CHCl_3$	0.01	—
Chromium VI	Cr	3×10^{-3}	8×10^{-6}
Dichloromethane	CH_2Cl_2	6×10^{-3}	0.6
Ethylene glycol	$C_2H_6O_2$	2	—
Formaldehyde	CH_2O	0.2	—
Hydrogen sulfide	H_2S	—	0.002
Isobutyl alcohol	$(CH_3)_2CHCH_2OH$	0.3	—
Mercury (elemental)	Hg	—	3×10^{-4}
Methanol	CH_3OH	2	20
Methylmercury	$[CH_3Hg]^+$	1×10^{-4}	—
Naphthalene	$C_{10}H_8$	0.02	3×10^{-3}
Nitrate	NO_3^-	1.6	—
Nitrite	NO_2^-	0.1	—
Phenol	C_6H_6O	0.3	—
Silver	Ag	5×10^{-3}	—
Sodium cyanide	NaCN	1×10^{-3}	—
1,1,1,2-Tetrachloroethane	$C_2H_2Cl_4$	0.03	—
1,1,2,2-Tetrachloroethane	$C_2H_2Cl_4$	0.02	—
Styrene	C_8H_8	0.2	1
Tetrachloroethylene ("Perc")	C_2Cl_4	6×10^{-3}	0.04
Toluene	C_7H_8	0.08	5
1,1,1-Trichloroethane	$C_2H_3Cl_3$	2	5
Vinyl chloride	C_2H_3Cl	3×10^{-3}	0.1
Warfarin		3×10^{-4}	—
Xylene (all forms)	C_8H_{10}	0.2	0.1

Source: [26] where values may be found for additional substances.

For inhalation, the reference dose is calculated from the reference concentration by the formula:

$$RfD \;=\; RfC\frac{CR}{BW} \,.$$

To assess the risk, one simply compares the exposure to the reference dose:

$$HQ = \frac{CDI}{RfD} \,.$$

The dimensionless ratio HQ stands for the *Hazard Quotient*, which is the measure of the risk. If $HQ < 1$, the risk is acceptable, but if $HQ > 1$, the risk is excessive.

10.4.2 Cancer risk

The calculation for cancer risk differs from that of noncancer risk because carcinogenic substances present a risk at all levels of exposure. There is no threshold below which there is no harm, and the risk of developing a cancerous tumor is the result of cumulative exposure. For most carcinogenic substances, the risk grows proportionally with exposure, and the key factor is the *Slope Factor SF* (sometimes also called the *potency factor*), which is the multiplier by which one determines the risk from the cumulated exposure, and the *Lifetime Cancer Risk* is calculated by the formula

$$Lifetime \; Cancer \; Risk = SF \times CDI,$$

in which the *CDI* was defined earlier and is expressed in mg/(kg. day). A lifetime cancer risk value of, say, 14×10^{-6} means that a person has 14 chances in a million to die of cancer due to that carcinogen in an otherwise expected lifetime of 70 years, or also that 14 people in a population of one million similarly exposed to the substance are expected to die of cancer sometime during their lifetime [27].

For some substances, the US Environmental Protection Agency prefers to list a *Unit Risk*, expressed in $(\mu g/L)^{-1}$ for oral ingestion or $(\mu g/m^3)^{-1}$ for inhalation, instead of the slope factor. In this case, one operates the conversion:

$$SF = \frac{BW}{CR} \; (Unit \; Risk).$$

Default values for the *Body Weight (BW)* and *Contact Rate (CR)* are given in a preceding table.

Slope Factors and *Unit Risks* for common carcinogens are listed in the following.

Slope factors unit risks for common carcinogens				
		Slope factor	Unit risk	
Carcinogen	Chemical formula	Ingested orally (mg/kg.day)$^{-1}$	Ingested orally (μg/L)$^{-1}$	Inhaled (μg/m^3)$^{-1}$
Arsenic	As	1.5	5×10^{-5}	4.3×10^{-3}
Benzene	C_6H_6	0.015	4.4×10^{-7}	2.2×10^{-6}
Carbon tetrachloride	CCl_4	0.07	2×10^{-6}	6×10^{-6}
Dichloromethane	CH_2Cl_2	2×10^{-3}	—	1×10^{-8}
Polychlorinated biphenyls	$C_{12}H_{10-x}Cl_x$	2	—	1×10^{-4}
1,1,1,2-Tetrachloroethane	$C_2H_2Cl_4$	0.026	7.4×10^{-7}	7.4×10^{-6}
1,1,2,2-Tetrachloroethane	$C_2H_2Cl_4$	0.2	6×10^{-6}	—
Tetrachloroethylene ("Perc")	C_2Cl_4	2.1×10^{-3}	6.1×10^{-8}	2.6×10^{-7}
1,1,2-Trichloroethane	$C_2H_3Cl_3$	0.057	1.6×10^{-6}	1.6×10^{-5}
Vinyl chloride	C_2H_3Cl	1.5	4.2×10^{-5}	8.8×10^{-6}

Source: [26] where values may be found for additional substances.

Sources

[1]   J. Xu, S.L. Murphy, K.D. Kochanek, E. Arias, Mortality in the United States, 2018. NCHS Data Brief. No. 355, 2020. January 2020, 8 pages. www.cdc.gov/nchs/data/databriefs/db355-h.pdf.

[2]   K.D. Kochanek, S.L. Murphy, J. Xu, E. Arias, Deaths: final data for 2017. National vital statistics reports, Cent. Dis. Control Prev. 68 (9) (2019), 77 pages. www.cdc.gov/nchs/data/nvsr/nvsr68/nvsr68_09-508.pdf.

[3]   U.S. Department of Transportation, National Highway Traffic Safety Administration - Traffic Deaths Decreased in 2018, but Still 36,560 People Died. www.nhtsa.gov/traffic-deaths-2018.

[4]   U.S. National Capital Poison Center, Poison Statistics — National Data, 2018. www.poison.org/poison-statistics-national.

[5]   U.S. Department of Justice, Federal Bureau of Investigation (FBI) 2018 Crime in the United States. ucr.fbi.gov/crime-in-the-u.s/2018/crime-in-the-u.s.-2018/topic-pages/murder.

[6]   Injury Facts, Odds of Dying. injuryfacts.nsc.org/all-injuries/preventable-death-overview/odds-of-dying/.

[7]   U.S. Department of Health and Human Services, Centers for Disease Control and Prevention — National Center for Health Statistics — AIDS and HIV. www.cdc.gov/nchs/fastats/aids-hiv.htm.

[8]   U.S. Department of Health and Human Services, Centers for Disease Control and Prevention — National Center for Health Statistics — First Data Released on Maternal Mortality in Over a Decade, posted 30 January 2020. www.cdc.gov/nchs/pressroom/nchs_press_releases/2020/202001_MMR.htm.

[9]   WordPress.com, History Lists — Human Deaths in the U.S. Caused by Animals. historylist.wordpress.com/2008/05/29/human-deaths-in-the-us-caused-by-animals/.

[10] The Economist, Air Safety — A crash course in probability, 29 January 2015. www.economist.com/gulliver/2015/01/29/a-crash-course-in-probability.

[11] U.S. Department of Health and Human Services, Centers for Disease Control and Prevention — Seasonal Influenze (Flu) — Disease Burden of Influenza. www.cdc.gov/flu/about/burden/index.html.
      Calculation was performed using 534,112 hospitalizations/year (average number of hospitalizations from 201415 to 2018-2019 flu seasons)

[12] U.S. Department of Health and Human Services, Centers for Disease Control and Prevention —Healthy Water — Current Waterborne Disease Burden Data & Gaps — 2017 Data. www.cdc.gov/healthywater/burden/current-data.html.

[13] Commonwealth of Massachusetts, Health and Human Services — Hospitalization-Epht-46-Issue-Test-Report (Form to Be Filled Out to Generate Report; Parameters Used: Asthma, Suffolk County, 2016-2016, Hospital Admission; Report Stated: 853 Hospitalizations in Population of 794,970, Adjusted Rate Per 10,000 Equal to 12.6). cognos10.hhs.state.ma.us/cv10pub/cgi-bin/cognosisapi.dll.

[14] Crime in America, The Probability of Being A Victim of Violent Crime, March 2, 2020. www.crimeinamerica.net/the-probability-of-being-a-victim-of-violent-crime/.

[15] U.S. Department of Health and Human Services, Centers for Disease Control and Prevention —Smoking & Tobacco Use.
      (1) Fast Facts at www.cdc.gov/tobacco/data_statistics/fact_sheets/fast_facts/index.htm
      (2) Current Cigarette Smoking Among Adults in the United States at www.cdc.gov/tobacco/data_statistics/fact_sheets/adult_data/cig_smoking/index.htm

[16] U.S. Department of Transportation, National Highway Traffic Safety Administration (NHTSA) — Traffic Safety Facts — 2017 Data — Motorcycles, DOT HS 812 785, August 2019, 10 pages. crashstats.nhtsa.dot.gov/Api/Public/ViewPublication/812785.

[17] U.S. Department of Health and Human Services, Centers for Disease Control and Prevention, The National Institute for Occupational Safety and Health (NIOSH), Agricultural Safety. www.cdc.gov/niosh/topics/aginjury/default.html.

[18] U.S. Department of labor, Mine Safety and Health Administration (MSHA) — Coal Fatalities for 1900 through 2019. arlweb.msha.gov/stats/centurystats/coalstats.asp.

[19]   Insurance Institute for Highway Safety (IIHS), Highway Loss Data Institute (HLDI) −
        Fatality Facts 2018, State by State, posted December 2019. www.iihs.org/topics/fatality-
        statistics/detail/state-by-state.

[20]   G.M. Masters, Introduction to Environmental Engineering and Science, second ed., Prentice
        Hall, 1997, 651 pages.

[21]   P.R. Hunter, L. Fewtrell, Acceptable Risk. Chapter 10 in Water Quality: Guidelines,
        Standards and Health, in: L. Fewtrell, J. Bartram (Eds.), World Health Organization (WHO),
        IWA Publishing, London, 2001, pp. 207−227. www.who.int/water_sanitation_health/dwq/
        iwachap10.pdf.

[22]   R. Wilson, Analyzing the daily risks of life, Technol. Rev. 81 (4) (1979) 41−46.

[23]   L. Laudan, The Book of Risks − Fascinating Facts about the Chances We Take Every Day,
        Wiley, 1994, 221 pages.

[24]   U.S. Environmental Protection Agency (EPA), Exposure Factors Handbook, Document
        EPA/600/R-09/052F, September 2011 with 2019 updates. www.epa.gov/expobox/about-
        exposure-factors-handbook.

[25]   U.S. Environmental Protection Agency (EPA), Integrated Risk Information System (IRIS) −
        Reference Dose (RfD): Description and Use in Health Risk Assessments, Background
        Document 1A, March 15, 1993. www.epa.gov/iris/reference-dose-rfd-description-and-use-
        health-risk-assessments.

[26]   U.S. Environmental Protection Agency (EPA), Integrated Risk Information System (IRIS) −
        IRIS Advanced Search. cfpub.epa.gov/ncea/iris/search/.

[27]   U.S. Environmental Protection Agency (EPA), Guidelines for Carcinogen Risk Assessment,
        Document EPA/630/P-03/001F, March 2005, 166 pages. www.epa.gov/sites/production/
        files/2013-09/documents/cancer_guidelines_final_3-25-05.pdf.

Chapter 11

Waste, packaging, and recycling

Much about waste and recycling depends on the level of economic activity, causing numbers such as total amounts of waste and recycling rates to fluctuate widely over time and across regions. As there is no benefit in cataloging here numbers that are highly variable, this chapter restricts the attention to the more stable numbers that characterize the various types and proportions of solid waste, their rate of decomposition, their carbon contents, and the comparative environmental benefits of incinerating, recycling, and composting per amount diverted from landfills.

11.1 Municipal solid waste

Municipal solid waste is the technical term used to refer to what is commonly called trash, rubbish or garbage, materials discarded by households and collected by municipalities and destined to a landfill, an incinerator, a recycling stream, or composting. It includes not only goods that have reached the end of their useful life, but also transient materials used in packaging and unwanted items such as yard waste, leftover food, and construction scrap.

In developed countries, municipal solid waste tends to have a consistent composition profile, as depicted in Figure 11-1.

In the United States, the per-capita rate of generation hovers around 4.51 lbs (2.05 kg) per person per day, with slight variation from year to year [2]. This means that a person weighing 170 lbs (77 kg) will throw away 755 times their own weight during their expected 78-year lifetime.

The discarded materials end up mostly in landfills (52.5%) with the rest being incinerated (12.8%), or diverted for recycling (25.7%) or composting (9.0%) [1,2].

The disposal of 1 metric ton of waste in a landfill necessitates the following inputs: 1.8 kg of diesel fuel, 8 kWh of electricity, 1 kg of plastic liner material (HDPE), and 100 kg of gravel ([3] Table 3).

Data, Statistics, and Useful Numbers for Environmental Sustainability.
https://doi.org/10.1016/B978-0-12-822958-3.00005-4
163

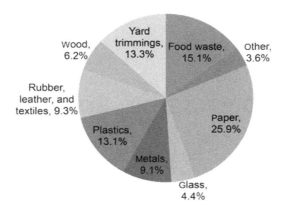

FIGURE 11-1 Typical composition of municipal solid waste on a per-mass basis. *Source:* [1].

As shown in Figure 11-1, the largest contributor by weight is paper, accounting for about 26% (or 1.2 lbs per person per day). This includes magazines and other paper-based items that usually enter the house through the mail as well as cardboard and other packaging material discarded immediately after purchase. With few exceptions, this material is recyclable and ought not to constitute such a large fraction of municipal solid waste. More on this in Section 11.2.

Next, in order of significance is food waste, accounting for about 15% (or 0.68 lb per person per day). Preferred alternatives are avoidance and composting. See Section 11.7.

Third in order of prevalence are miscellaneous types of plastics, originating mostly from packaging and disposable items like plastic shopping bags, beverage bottles, disposable silverware, and toys. These account for about 13% of all solid waste or 0.59 lb per person per day. The problem with plastics is that they come in many different types requiring sorting before recycling. Many of the plastics are parts of larger items such as covers of small appliances for which recycling can only proceed after disassembly. For unsorted plastics, incineration (with proper precautions) is the preferred alternative. A detailed list of plastics, their embodied energy, their carbon footprint, and water use during production can be found in Section 1.2. Some of their decomposition times in landfills are listed in the following.

Decomposition times of different materials in a landfill

Item	Decomposition time	Item	Decomposition time
Paper towel	2 to 4 weeks	Wool clothing	1 to 5 years
Banana or orange peel	2 to 5 weeks	Plastic shopping bag	10 to 20 years
Paper bag	1 month	Nylon fabric	30 to 40 years
Paper	2 to 6 weeks	Leather	50 years
Apple core	2 months	Tin can	50 years
Cardboard	2 months	Aluminum can	80 to 200 years
Cotton T-shirt	2 to 5 months	Plastic bottle	450 years
Waxed milk carton	3 months	Glass bottle	1 million years
Plywood	1 to 3 years	Styrofoam	Does not biodegrade

Sources: [4,5].

11.2 Packaging

Corrugated cardboard as that found in cardboard boxes used in shipping is imminently recyclable. Of those used for bulk deliveries in the rear of retail stores, 95% are captured and recycled, with the new boxes containing 38% recycled fibers and 62% new fibers for strength ([6] page 1–16). The recycling of cardboard boxes used by *e*-commerce sellers depends on the willingness of the customer, but they are generally recycled to a high degree.

An alternative to the cardboard box is the hard plastic crate with a hinged cover. Such a crate typically weighs 4.25 lbs (1.93 kg), which is 2.3 times heavier than the cardboard box it replaces ([6] page ES-3), but can be used an average of 200 times over the course of 10–15 years ([7] page 4). The analysis of several applications in transportation and display of fresh produce revealed that reusable crates, including their transportation and washing between uses, require 39% less total energy, produce 95% less total solid waste, and generate 29% less total greenhouse gas emissions than the 38%-recycled-content cardboard box ([6] page ES-1).

Packaging of beverages is generally in the form of plastic bottles, glass bottles, or aluminum cans. The following table compares the energy, solid waste, and carbon footprints of the three types of containers.

Impacts of beverage containers per 1,000 L of beverage				
		Solid waste		
Container	Energy (MJ)	Weight (kg)	Volume (m^3)	Greenhouse gases (kg CO$_{2eq}$)
PET plastic bottle	4,080	48.2	0.18	180
Glass bottle	9,880	711.5	0.58	774
Aluminum can	5,940	122.4	0.26	442

Source: [8] citing an independent study by Franklin Associates, with numbers converted to metric units.

11.3 Electrical and electronic equipment

End-of-life electrical and electronic objects go by the technical name of Waste Electrical and Electronic Equipment (WEEE) and more colloquially by the expression *e*-waste. Their typical makeup in the waste stream is as tabulated in the following, first by object type and then by material kind and small parts.

Typical composition of WEEE by object type	
Equipment type	**Weight fraction**
Large household appliances	49.1%
Small household appliances	7.0%
Phones, laptops, and other IT devices	16.3%
Consumer electronics (TVs, DVD players, etc.)	21.1%
Lighting equipment (lamps)	2.40%
Electric tools	3.52%
Control instruments (thermostats, detectors)	0.21%
Automatic dispensers (for cash, drinks, etc.)	0.18%
Medical devices	0.12%
Toys, leisure and sports equipment	0.11%
Source: [9] Table 1, representative of WEEE in the European Union.	

Typical composition of WEEE by material	
Materials or small parts	**Weight fraction**
Metals	60%
Plastics	15%
Metal-plastic mixtures	5%
Remainders of CRT & LCD screens	12%
Printed circuit boards	2%
Cables	2%
Various pollutants	3%
Other	1%
Source: [9] Fig. 1.	

Because they contain toxic substances such as lead, mercury, arsenic, and brominated flame retardants in circuit boards and casings, computers and their monitors should not be landfilled. Doing so would result in toxic leakages to the groundwater. Samples taken in both the United Kingdom and the United States have revealed lead concentration in water of about 0.13 mg/L with high values reaching 0.250 mg/L ([10] Section 2). When e-waste is shipped to developing nations for recycling, the informal methods of disassembly, with some using open-pit melting components for material separation, cause toxic emissions with lead concentration reaching 444 ng/m^3 in the surrounding air ([10] Table 2).

The overall recycling rate of smartphones in Europe is estimated at 83% with the remaining 17% going to the landfill [11].

11.4 Construction waste

In the United States, construction and demolition debris end up in municipal landfills, where they account for about 24% of the mass. The wasted materials include wood (27% of construction/demolition waste) as well as masonry, bricks, concrete, drywall/plaster, dirt rubble, insulation, siding, roofing, asphalt, waterproofing material, metal, cardboard, paper, and landscaping material ([12] Section 1.4.12).

The following table illustrates the typical amounts of wasted materials generated by the construction of a 2,000-ft^2 private house in the United States.

Amounts of construction waste by material type					
	Weight			Volume	
Material	lbs	kg	Fraction	ft^3	m^3
Solid sawn wood	1,600	726	20%	160	4.6
Engineered wood boards	1,400	635	18%	130	3.8
Drywall[1]	2,000	910	25%	160	4.6
Cardboard	600	270	7.5%	540	15.2
Metals	150	68	1.9%	30	0.8
Plastics (PVC)	150	68	1.9%	30	0.8
Masonry	1,000	450	13%	30	0.8
Hazardous materials[2]	50	23	0.6%	–	–
Other	1,050	480	13%	300	8.4
TOTAL	8,000	3,630	100%	1,380	39

[1]*Drywall, also called gypsum wallboard, consists of a sheet of plaster sandwiched between two layers of cardboard; it is used for interior walls.*
[2]*Hazardous materials include paint, caulk, roofing cement, aerosols, solvents, adhesives, oils, and greases.*
Source: [12] Section 1.4.13.

11.5 Recycling

Recycling rates are not systematically mentioned here as they vary widely across countries and over time due to highly fluctuating profit margins (or absence thereof). Instead, this section focuses on the benefits of recycling per unit mass of material recycled.

The recycling of 1 kg of steel saves 1.25 kg of iron ore, 0.70 kg of coal, and 0.06 kg of limestone [13].

It is estimated that the reading of the New York Times daily, including on Sundays when the print edition is significantly thicker, amounts to 207 kg (455 lbs) of CO_{2eq} per year if the papers are all recycled or to 447 kg (984 lbs) of CO_{2eq} if they are thrown in the garbage bin ([14] pages 48−49). The latter is nearly equivalent to a roundtrip flight between New York and Atlanta (487 kg of CO_{2eq}).

Almost all (estimated 95%) of construction/demolition waste is clean, unmixed, and recyclable ([12] Section 1.4.12). Some data indicate that about 48% is being separated for recycling ([12] Section 1.14.15)

11.5.1 Energy savings from recycling

The production of many materials is lower when recycled than procured from virgin sources. The difference is greatest for aluminum, for which the production of 1 kg necessitates 210 MJ from bauxite and only 26 MJ from a recycled beverage can ([15] page 471), a saving of 184 MJ/kg or 88%. For a single 15-g aluminum can, its recycling translates into conserving the energy to power a 60-W light bulb for 13 hours. For low carbon steel, the corresponding numbers are 26.5 MJ/kg from iron ore versus 7.3 MJ/kg from recycling ([15] page 463), a saving of 19.2 MJ/kg or 72%.

The following table lists the energy saved when a material is diverted from the landfill to a recycling stream. Note that this is not exactly the same energy savings as when production is switched from virgin to recycled sources because the recycling stream may lead the material into secondary products, like automobile steel into washing machines. Also, postconsumer paper diverted from the landfill must be transported to a more distant paper mill, subtracting from the energy saving.

Energy savings from recycling various materials		
	Energy savings	
Material	10^6 BTUs per short ton	MJ per kg
Aluminum cans[1]	206.9	240.6
Steel cans	20.5	23.8
Copper wire	83.1	96.7
Glass	2.65	3.08
Plastic—HDPE	51.4	59.8
Plastic—LDPE	56.5	65.7

Continued

Plastic—PET	53.4	62.1
Corrugated cardboard	15.7	18.3
Newspaper	16.9	19.7
Textbooks	0.7	0.8
White office paper	10.2	11.9
Carpets	106.1	123.4

[1]*The number quoted here (240.6 MJ/kg) differs from the difference quoted in the preceding text (184 MJ/kg) because the production of aluminum from bauxite necessitates multiple intercontinental transports whereas production of aluminum from beverage cans only involves regional transport. Hence, the savings are larger when transportation is considered.*
Source: [16] Exhibit 7-1.

11.5.2 Greenhouse gas and other savings from recycling

Diverting 1 short ton (=907 kg) of glass from the landfill to recycling saves 2.65 million BTUs (=3.08 MJ/kg) and 90 kg of CO_{2eq} (= 99 g CO_{2eq}/kg) ([16] page 100).

There are significant multiple benefits to 100% recycled office paper compared to paper made from 100% virgin fiber, as shown in the following.

Benefits of recycling white office paper	
Impact	**Reduction**
Total energy consumption	44%
Net greenhouse gas emissions	38%
Particulate emissions	41%
Solid waste	49%
Wastewater	50%
Wood use	100%

Source: [17] page 14.

For one ream (500 sheets weighing 5 lbs = 2.27 kg) of office paper with 30% post-consumer recycled content, this translates into a savings of 14,600 BTUs (15.4 MJ) of energy, 0.66 lbs (300 g) of CO_{2eq}, and 2.2 gallons (8.5 L) of wastewater.

Proceeding from life-cycle assessments, one can calculate so-called emission factors that estimate, on a per-mass basis, the greenhouse gas emissions of the recycling process and the net carbon savings after accounting for the avoidance of virgin material and associated energy consumption. The following table provides a long list of emission factors. Negative values indicate greenhouse gas emission savings.

Carbon emission factors for the recycling of various materials and objects

Material	Recycling process emissions (kg CO₂eq/kg)	Net after accounting for virgin material avoidance (kg CO₂eq/kg)	Material	Recycling process emissions (kg CO₂eq/kg)	Net after accounting for virgin material avoidance (kg CO₂eq/kg)
Aluminum cans	1,113	−8,143	Paint	364	+86
Aluminum foil	1,113	−8,143	Paper—books	562	−117
Appliances—small	463	−1,349	Paper—white	1,576	−459
Appliances—large	428	−866	Paper—mixed	559	−120
Batteries—auto	938	−435	Plasterboard	59	+4
Batteries—other	1,129	−205	Plastics—HDPE	379	−1,149
Bicycles	883	−3,577	Plastics—LDPE	29	−972
Cardboard	559	−120	Plastics—PET	155	−2,192
Carpets	181	−10	Plastics—PP	379	−1,184
CRT display screens	272	−228	Plastics—PVC	379	−1,549
Food cartons	629	−452	Plastics—mixed	339	−1,024
Fire extinguishers	651	−673	Refrigerators	469	−853
Fluorescent tubes	518	−779	Spray cans	883	−3,577
Footwear	401	−3,376	Steel cans	529	−862

Continued

Carbon emission factors for the recycling of various materials and objects—cont'd

Furniture	502	−444	
Glass	395	−314	
Light bulbs	518	−779	
Mattresses	478	−1,241	
Metal—scrap	883	−3,577	
Oil—mineral	647	−2,759	
Oil—vegetable	647	−2,759	
Textiles	401	−3,376	
Tires—cars	206	−636	
Tires—trucks	197	−671	
Tires—mixed	206	−636	
Wood—solid	502	−444	
Wood—composite	502	−444	

Source: [3] Table 6.

11.6 Methane capture

On a dry weight basis, the municipal solid waste contains 30−50% cellulose, 7−12% hemicellulose, and 15−18% lignin ([16] page 82). The carbon content of these biodegrades (lignin to a lesser extent than the other two), and in the anoxic environment of a landfill this decomposition produces methane gas (CH_4). As methane is a hydrocarbon that can be used as a fuel, there is a benefit to capturing the methane emanating from landfills. The following table gives the amounts that can be expected given favorable conditions and sufficient time for complete decomposition. When not captured, the methane contributes to greenhouse gas emissions (with 1 kg methane counting as equivalent to 32 kg of CO_2), from which the contribution of untapped landfills to climate change may be estimated.

Methane emissions from landfills				
Material	Initial carbon content	Methane yield (metric ton of C/wet short ton)	Methane yield (kg CH_4/kg)	CO_{2eq} emission if CH_4 not captured (kg CO_{2eq}/kg)
Corrugated cardboard	47%	0.688	1.01	32.4
Magazines and junk mail	34%	0.278	0.409	13.1
Newspaper	49%	0.244	0.359	11.5
White office paper	40%	1.198	1.761	56.3
Food waste	51%	0.445	0.654	20.9
Yard trimmings — Grass	45%	0.150	0.220	7.05
Yard trimmings — Leaves	41%	0.281	0.413	13.2
Yard trimmings — Branches	49%	0.355	0.522	16.7
Mixed municipal solid waste	42%	0.580	0.852	27.3

Source: [16] Exhibit 6-3 based on data reported in [18].

The US Environmental Protection Agency [19] mentions the following benefits of methane gas capture from landfills with subsequent electricity generation.

(1) A typical direct-use project using 1,000 ft^3 of landfill gas per minute is equivalent to the annual sequestration of about 163,000 acres (=255 square miles) of forests, also equivalent to the annual avoidance of CO_2 emissions from more than 322,000 barrels of oil, or from nearly 15.6 million gallons (=59 million L) of gasoline, or from heating about 3,900 homes.

(2) A typical 3-MW electricity generation plant reduces greenhouse gas emissions equivalent to the sequestration of about 178,000 acres (=278 square miles) of forests in 1 year, also equivalent to the avoidance of CO_2 emissions from the consumption of 17.0 million gallons (=64 million L) of gasoline, or from the burning of the coal from nearly 830 railcars, or from the electricity needed to power about 1,900 homes.

For more equivalencies, use the EPA's Greenhouse Gas Equivalencies Calculator [20].

A useful tool for the estimation of greenhouse gas emissions from a specific landfill is the Waste Reduction Model created by the US Environmental Protection Agency [21]. For example, The State of Vermont (USA) used the model to determine that the methane gas reduction resulting from its diversion of 80,000 short tons/year of food waste from landfills to composting is equivalent to removing 7,000 passenger cars from the road [22].

11.7 Composting

Aside from its advantage of diverting solid waste from landfills, composting of organic material generates heat (see earlier Section 3.3.4) starting at a rate of 1,500 BTUs per hour per short ton (=0.48 W/kg) for the first 4 weeks and decreasing to a third of that for the next 4 weeks ([23] page 118). Cumulated over an 18-month period, the heat generation amounts to 6.4 million BTUs per short ton (=7.4 MJ/kg) ([23] page 120). Other sources estimate the averaged heat recovery from a compost in a commercial system to be slightly lower, at 7.1 MJ per kg of dry matter [24].

The carbon dioxide emission from a compost pile is of no concern to climate change as this is not new carbon from a fossil fuel.

11.8 Waste in the ocean

The most common kinds of waste found in the ocean are food wrappers, beverage bottles, grocery bags, straws, and take-out containers, all made of plastic. Additionally, there is discarded fishing gear, mostly made of nylon fibers. Neither plastics nor nylon fibers biodegrade in the ocean; they only break down into minute pieces shorter than 5 mm (0.2 in) called microplastics. Some microplastics also originate from microbeads found in products such as toothpaste and face wash. It is estimated that 8 million metric tons of plastic enter the ocean every year, the weight equivalent of nearly 90 aircraft carriers [25].

In the North Pacific Ocean between 35–42°N and 135–155°W, an ocean gyre has trapped an abundance of plastics in an area now dubbed the *Great Pacific Garbage Patch*. More than 75% of the mass consists of debris larger than 5 cm, with at least 46% originating from fishing nets. Microplastics account for only 8% of the total mass but 94% of the estimated 1.8 trillion pieces floating in the area. The area is approximately 1.6 million km^2 (620,000 square miles) [26]. This is more than twice the size of Texas (269,000 square miles).

A Danish research project estimates that microplastics are now finding their way into 16 out of 17 brands of sea salt, 4 out of 5 samples of drinking water, and 80% of British mussels. Effects on humans are unknown at the time of this writing, but there is mounting evidence of harm to animal life [27].

The following table provides estimates of the origins of microplastics.

Microplastics in the ocean	
Origin	**Weight fraction**
Synthetic textiles	35%
Synthetic rubber (tires)	28%
City dust and debris	24%
Road markings	7%
Marine coatings	4%
Care products with microbeads	2%
Plastic pellets	0.3%

Source: [28], see also an illustration in [27] but with some incorrect numbers.

Sources

[1] University of Michigan, Center for Sustainable Systems — Municipal Solid Waste Factsheet, Pub. No. CSS04-15, 2019. css.umich.edu/factsheets/municipal-solid-waste-factsheet.

[2] U.S. Environmental protection Agency, Facts and Figures about Materials, Waste and Recycling. www.epa.gov/facts-and-figures-about-materials-waste-and-recycling/national-overview-facts-and-figures-materials.

See also: Municipal Solid Waste Generation, Recycling, and Disposal in the United States: Facts and Figures. archive.epa.gov/epawaste/nonhaz/municipal/web/html/msw99.html.

[3] D.A. Turner, I.D. Williams, S. Kemp, Greenhouse gas emission factors for recycling of source-segregated waste materials, Res. Conserv. Recycl. 105, 2015, 186—197, doi.org/10.1016/j.resconrec.2015.10.026.

[4] Save On Energy, Material Decomposition — How Long it Takes for Trash to Decompose. www.saveonenergy.com/material-decomposition/.

[5] Keep Cass County Beautiful — Recycling — How Long Does It Take Garbage to Decompose? www.keepcasscountybeautiful.com/images/PDF/Recycling/how_long_does_it_take_garbage_to_decompose.pdf.

[6] Franklin Associates, Life Cycle Inventory of Reusable Plastic Containers and Display-Ready Corrugated Containers Used for Fresh Produce Applications. Report Prepared for the Reusable Pallet & Container Coalition, 2004. November 2004, 80 pages, reusables.org/wp-content/uploads/2016/06/FinalRprt-FranklinLCI-nov04.pdf.

[7] Svenska Retursystem, Eurocrate: A full demonstration of reusable crates and pallets. Undated, 8 pages. ec.europa.eu/environment/life/project/Projects/index.cfm?fuseaction=home.showFile &rep=file&fil=LIFE00_ENV_S_000867_LAYMAN.pdf.

[8] PET Resin Association (PETRA), News and Information − New Study Gives "Green" Light to PET Bottles over Glass or Aluminum, dated 6 April 2010. petresin.org/news_GreenLighttoPETBottles.asp.

[9] F.O. Ongondo, I.D. Williams, T.J. Cherrett, How are WEEE doing? A global review of the management of electrical and electronic wastes, Waste Manag. 31 (4), 2011, 714−730.

[10] E. Williams, R. Kahhat, B. Allenby, E. Kavazanjian, J. Kim, M. Xu, Environmental, social, and economic implications of global reuse and recycling of personal computers, Environ. Sci. Technol. 42 (17), 2008, 6446−6454, doi.org/10.1021/es702255z.

[11] M. Ercan, J. Malmodin, P. Bergmark, E. Kimfalk, E. Nilsson, Life cycle assessment of a smartphone, in: Advances in Computer Science Research, Proceedings of 4th Int. Conf. on ICT for Sustainability (ICT4S 2016), 2016, pp. 124−133, doi.org/10.2991/ict4s-16.2016.15.

[12] U.S. Department of Energy, Energy Efficiency & Renewable Energy − 2011 Buildings Energy Data Book, 286 pages. ieer.org/wp/wp-content/uploads/2012/03/DOE-2011-Buildings-Energy-DataBook-BEDB.pdf.

[13] American Iron and Steel Institute, Steel Recycling Institute − Steel: The Natural Choice for Buying Recycled, dated 2020. www.steelsustainability.org/recycling/buy-recycled.

[14] M. Berners-Lee, How Bad Are Bananas? The Carbon Footprint of Everything, Greystone Books, D&M Publishers, 2011, 232 pages.

[15] M.F. Ashby, Materials and the Environment − Eco-Informed Material Choice, second ed., Butterworth-Heinemann, 2013, 616 pages.

[16] U.S. Environmental Protection Agency, National Service Center for Environmental Publications (NSCEP): Solid Waste Management and Greenhouse Gases: A Life-Cycle Assessment of Emissions and Sinks, third ed., (September 2006), 142 pages. nepis.epa.gov/Exe/ and search by title. Also available from: large.stanford.edu/courses/2011/ph240/machala2/docs/fullreport.pdf.

[17] J. Roberts (Ed.), The State of the Paper Industry − Monitoring the Indicators of Environmental Performance, Report by The Environmental Paper Network, 2007, 77 pages.

[18] M.A. Barlaz, R.K. Ham, D.M. Schaefer, Methane production from municipal refuse: a review of enhancement techniques and microbial dynamics, Crit. Rev. Environ. Control 19 (6), 1990, 557−584, doi.org/10.1080/10643389009388384.

[19] U.S. Environmental Protection Agency, Landfill Methane Outreach Program (LMOP) − Landfill Gas Energy Benefits Calculator. www.epa.gov/lmop/landfill-gas-energy-benefits-calculator.

[20] U.S. Environmental Protection Agency, Energy and the Environment − Greenhouse Gas Equivalencies Calculator. www.epa.gov/energy/greenhouse-gas-equivalencies-calculator.

[21] U.S. Environmental Protection Agency, Documentation Chapters for Greenhouse Gas Emission, Energy and Economic Factors Used in the Waste Reduction Model (WARM) − Electronics Chapter, 23 pages. www.epa.gov/sites/production/files/2019-06/documents/warm_v15_electronics.pdf.

The accompanying Excel™ calculator can be downloaded from: www.epa.gov/warm/versions-waste-reduction-model-warm#15.

[22]   Vermont Public Radio — Vermont Edition — How Composting Fits In To Vermont's Recycling Conversation, conversation by host Jane Lindholm with Josh Kelly of the Solid Waste program of the Vermont Agency of Natural Resources in the Department of Environmental Conservation, October 8, 2019. www.vpr.org/post/how-composting-fits-vermonts-recycling-conversation.

[23]   G. Brown, The Compost-Powered Water Heater, The Countryman Press, Woodstock Vermont, 2014, 162 pages.

[24]   M.M. Smith, J.D. Aber, R. Rynk, Heat recovery from composting: a comprehensive review of system design, recovery rate, and utilization, Compost Sci. Util. 25 (2017) S21—S22, doi.org/10.1080/1065657X.2016.1233082.

[25]   U.S. Department of Commerce, National Oceanic and Atmospheric Administration — National Ocean Service — Hazards — A Guide to Plastic in the Ocean, undated. oceanservice.noaa.gov/hazards/marinedebris/plastics-in-the-ocean.html.

[26]   L. Lebreton, B. Slat, F. Ferrari, B. Sainte-Rose, J. Aitken, R. Marthouse, S. Hajbane, S. Cunsolo, A. Schwarz, A. Levivier, K. Noble, P. Debeljak, H. Maral, R. Schoeneich-Argent, R. Brambini, J. Reisser, Evidence that the Great Pacific garbage patch is rapidly accumulating plastic, Nature Sci. Rep. 8 (2018) 4666, doi.org/10.1038/s41598-018-22939-w, 15 pages.

[27]   R. Orange, How to Clean Microplastics from the Ocean, Alfa Laval Corporate AB, Lund, Sweden, 2019 posted under the rubric "Media — Stories" on 25 January 2019, www.alfalaval.com/media/stories/municipal-wastewater-treatment/membranes-a-solution-to-microplastics-in-our-oceans/.

[28]   HORIBA — Science in Action Series, Where do microplastics come from? Posting on, July 29, 2020. www.horiba.com/en_en/science-in-action/where-do-microplastics-come-from/.

Chapter 12

Industries

The spectrum of industries varies from country to country. As an indication of the impacts in a highly developed economy, the breakdown of energy consumption and carbon footprint in the United Kingdom and of water in California is as tabulated in the following.

Energy, carbon, and water footprints across the industrial system of a highly developed economy			
Sector	Energy (PJ/year)	Greenhouse gas emissions	Water
Chemicals	220	19%	(with "others")
Food and drink	150	7%	21%
Metals	110	25% steel 4% aluminum	11%
Rubber and plastics	95	4% plastics	(with "others")
Mineral products	95	(with "others")	(with "others")
Pulp and paper	70	6%	5%
Motor vehicles	60	3%	(with "others")
Electronics	40	(with "others")	13%
Cement	(not stated)	8%	(with "others")
Textiles	(not stated)	(with "others")	6%
Others not counted above	(not stated)	24%	44%

Sources: [1] Figs. 1 (for energy) and 3 (for greenhouse gases), [2] Fig. 4-2 (for water).

Data, Statistics, and Useful Numbers for Environmental Sustainability.
https://doi.org/10.1016/B978-0-12-822958-3.00006-6

12.1 Agriculture, livestock, and food industry

12.1.1 Land and water use

Needless to say, agriculture accounts for the largest share of land used by humans. Agriculture alone occupies 36% of the world's total land area, which is 50% of the habitable land, amounting (in 2013) to 0.70 hectares (1.7 acres) per person [3].

Breakdown of the earth's surface by type and use							
Oceans and seas	71%						
Land surface	29%	Habitable land	71%	Agriculture	50%	Livestock and feed	77%
						Crops	23%
				Forests	37%		
				Shrub	11%		
				Lakes, rivers	1%		
				Built	1%		
		Barren land	19%				
		Glaciers	10%				

Source: [3] Numbers from chart titled *Global land use for food production.*

Agriculture is responsible for 92% of water consumption by humans [4].

Of the several sectors comprising agriculture, the livestock sector (meat and dairy production) is particularly impacting. It uses about three quarters (77%) of all agricultural land, compared to 23% devoted to crops [3]. Emissions from livestock operations account for 37% of all human-induced methane, most from the digestive system of ruminants, and 64% of ammonia, which contributes to acid rain [5]. Also, 33% of the global arable land is dedicated to the production of feed for livestock [5].

12.1.2 Production efficiency and impacts

The carbon footprint of fertilizer is 5.6 kg CO_{2eq} per kg of applied nitrogen (N) [6]. The World Bank estimates the world's use of fertilizer at 141 kg/ha (=125 lbs/acre) [7]. The nitrogen content in fertilizers varies and is on the order of one-third by weight.

Agriculture is responsible for 95% of ammonia (NH_3) emissions [8].

The production of animal products is not efficient as the next two tables show. After animal products, the most land-intensive crops are chocolate, coffee, and fruits.

Feed conversion efficiency	
Animal	**kg dry feed/kg output**
Beef cattle	46.9
Dairy cattle	1.9
Sheep and goat	30.2
Pig	5.8
Broiler chicken	4.2
Layer chicken	3.1
Eggs	2.3
Milk	0.70
Source: [9] Table 1, [10].	

Land use per 100 g of protein produced	
Sheep and lamb	184.8 m^2
Beef cattle	163.6 m^2
Chocolate	137.9 m^2
Cheese	39.8 m^2
Milk	27.1 m^2
Coffee	27.0 m^2
Bananas	21.4 m^2
Apples	21.0 m^2
Citrus fruits	14.3 m^2
Pork meat	10.7 m^2
Nuts	7.9 m^2
Tomatoes	7.3 m^2
Poultry meat	7.1 m^2
Eggs	5.7 m^2
Potatoes	5.2 m^2
Grains	4.6 m^2

Continued

Rice	3.9 m^2
Fish (farmed)	3.7 m^2
Groundnuts (ex. peanuts)	3.5 m^2
Root vegetables	3.3 m^2
Tofu (soybeans)	2.2 m^2

Source: [3] Chart titled *Land use per 100 g of protein.*

The following table provides the yields and estimated greenhouse gas emissions of major crops.

		\multicolumn{2}{c}{CO_{2eq} emissions}
Product	**Yield**	**per mass (kg/kg)**	**per calorie (g/kcal)**
Beef meat	222.9 kg/animal	22.1	10.8
Cereals	4.07 metric ton/hectare	2.97	1.48
Cocoa beans	0.44 metric ton/hectare	−	−
Coffee beans	0.97 metric ton/hectare	−	−
Eggs	10.76 kg/bird	1.6	1.07
Fibers (vegetal origin)	0.84 metric ton/hectare	−	−
Forage crops[1]	13.7 metric ton/hectare	−	−
Fruits	11.3 metric ton/hectare	1.6	4.1
Lamb		39	14
Milk	1,039.7 kg/animal	2.4	5.7
Nuts	1.50 metric ton/hectare	−	−
Palm oil	14.38 metric ton/hectare	−	−
Pork meat	79.7 kg/animal	6.1	2.5
Potatoes	20.95 metric ton/hectare	3	4
Poultry meat	1.71 kg/animal	5.3	2.21
Pulses[2]	0.90 metric ton/hectare	−	−
Rice	3.71 metric ton/hectare	−	−
Roots and tubers[3]	15.0 metric ton/hectare	0.5	−
Sheep and goat meat	14.3 kg/animal	24	8.2

Table title: Yield and carbon footprint of various agricultural products

Continued

Sugar cane	72.6 metric ton/hectare	–	–
Tea	1.51 metric ton/hectare	–	–
Vegetables	19.5 metric ton/hectare	1.6	4.1

[1]*Permanent crops that are cultivated primarily for animal feed.*
[2]*Annual leguminous crops yielding grains or seeds used for food, feed, and sowing purposes and limited to crops harvested for dry grain only. Excluding: crops harvested green for forage and food, those used mainly for extraction of oil, and leguminous crops whose seeds are used exclusively for sowing purposes.*
[3]*Excluding: those crops that are cultivated mainly for feed, for processing into sugar, or which are generally classified as "roots, bulb and tuberous vegetables."*
Sources: [10, 11], [12] Fig. 2, [13–17], and a web calorie calculator.

12.1.3 Water footprints

The water consumed in the production of major crops is tabulated in the following. [*Note*: Green water is water from precipitation; blue water is fresh surface water or groundwater; gray water is the water required downstream to assimilate the pollutant load.]

Water footprint of various agricultural products				
Product	Green (m³/ton)	Blue (m³/ton)	Grey (m³/ton)	Total (m³/ton)	Per calorie (L/kcal)
Beer	–	–	–	300	–
Bovine meat	14,414	550	451	15,415	10.19
Butter	4,695	465	393	5,553	0.72
Cereals	1,232	228	184	1,644	0.51
Eggs	2,592	244	429	3,265	2.29
Fibers (vegetal origin)	3,375	163	300	3,837	–
Fodder crops	207	27	20	253	–
Fruits	727	147	93	967	2.10
Milk	863	86	72	1,020	1.82
Nuts	7,016	1,367	680	9,063	3.63
Oil crops (vegetal)	2,023	220	121	2,364	0.81
Pork meat	4,907	459	622	5,988	2.15
Poultry meat	3,545	313	467	4,325	3.00
Pulses	3,180	141	734	4,055	1.19

Continued

Roots and tubers	327	16	43	387	0.47
Rubber	12,964	361	422	13,748	–
Sheep and goat meat	8,253	457	53	8,763	4.25
Spices	5,872	744	432	7,048	2.35
Sugar crops	130	52	15	197	0.68
Tobacco	2,021	205	700	2,925	–
Vegetables	194	43	85	322	1.34
Wine	450 ± 23 L/bottle	128 ± 7 L/bottle		870	–

Sources: [9] Table 6, [18] Table 2 and page 1583, [19] page 187.

12.1.4 Specific foods

The water footprint of coffee (130 L/cup for 7 g of roasted coffee/cup) is much larger than that of tea (27 L/cup for 3 g of black tea/cup) ([18] page 1583). For both, the water used in production far exceeds the amount of water in the beverage (about 120 mL/cup).

The following table lists the energy consumption, carbon footprint, and other impacts of end food. Food processing, packaging, transportation, refrigeration, and cooking/baking are included in these numbers. The source quoted for the tabulated numbers includes the breakdowns. The list is very incomplete, and some listed items do not have numbers in every column; this is because the set of life-cycle assessments of food items is very incomplete.

Footprints of food served				
Food group	**Food item**	**Energy demand (MJ/kg)**	**Carbon footprint[1] (kg CO_{2eq}/kg)**	**Water eutrophication**
Basic carbohydrate foods	Bread	11.7	0.98	155 g BOD/kg
	Pasta[2]	19.6	–	–
	Potatoes[3]	4.5	0.215	1.1 g PO_4^{3-}/kg
	Rice	9.8	6.4	–
Fruits	Apples (domestic)	5.95	0.55	–
	Bananas (imported)	–	0.48	–
	Oranges	3.8	0.3–0.7	–

Continued

Vegetables	Carrots (fresh)	2.1	0.733	—
	Tomatoes (open field)	1.5−5	0.081	—
	Tomatoes (greenhouse)	125−137	9.4	—
Dairy	Butter	32.2	12.1	—
	Cheese	54.0	8.8	—
	Eggs	14	5.53	77 g PO_4^{3-}/kg
	Ice cream	4.4	0.97	8 g PO_4^{3-}/kg
	Milk	8.43 MJ/L	1.06 g/L	6.3 g PO_4^{3-}/L
	Yogurt (flavored)	25.8 MJ/L	—	—
Meats	Beef	43.6	15−32	242 g PO_4^{3-}/kg
	Lamb	33	17.4	200 g PO_4^{3-}/kg
	Pork	27	5.0	100 g PO_4^{3-}/kg
	Poultry	15	4.57	49 g PO_4^{3-}/kg
Fish	Cod (frozen fillet)	60	3.96	—
	Salmon (farmed)	57.5	—	—

[1]The electricity mix used in the carbon estimations corresponds to that in Europe (around 0.4 kg CO_{2eq}/kWh).
[2]For fresh pasta, the energy demand is slightly higher.
[3]If the end-use is "French fries" ("chips" in the United Kingdom), energy demand increases by 2.7 MJ/kg.
Sources: [20−24].

The carbon footprint of butter is 3.7 times higher than that of margarine (12.1 vs. 3.3 kg CO_{2eq}/kg), mostly because of methane emissions by dairy cows [24].

12.1.5 Wine industry

In 2015, the global wine industry cultivated 7.5 million hectares of vines (∼0.5% of all arable lands), from which it produced 75.7 million metric tons of grapes yielding 27.4 million m^3 of wine [19]. This amounts to 1.01 kg of grapes and 0.37 L of wine per m^2 of the vineyard.

The carbon footprint of a 0.75 L-bottle of wine breaks down as tabulated in the following.

Carbon footprint of a bottle of wine		
Stage	**Carbon footprint (kg CO_{2eq}/bottle)**	
Vineyard planting	0.07 ± 0.12	
Viticulture and grape growing	0.38 ± 0.31	1.09 ± 0.83 for red wine
Winemaking	0.26 ± 0.33	1.36 ± 0.66 for white wine
Packaging (bottles, labels, and crates)	0.47 ± 0.24	
Distribution (transport and warehousing)	0.25 ± 0.29	
Storage and consumption	0.26 ± 0.29	
End-of-life	0.48 ± 1.02	
Total	2.17 ± 1.34	
Source: [19] Fig. 7.3.		

The most water consumptive stages are packaging and distribution accounting for 56 and 41%, respectively. After discounting for natural rainwater ("green water"), the overall water footprint of winemaking is nearly proportional to its carbon footprint, with a coefficient of proportionality equal to 81.24 L of water per kg of CO_{2eq} emitted [19].

Carbon footprint of a liter of alcohol		
Inputs	**Carbon footprint (kg CO_{2eq}/L)**	
Liquid fuels	0.807	
Propane	2.200	
Fertilizers	0.403	
Pesticides	0.110	3.74
Electricity	0.037	
Glass	0.183	
Freight transportation	5.44	
Total	9.18	
Source: [25] Fig. 4.		

The carbon footprint of cognac, a proxy for most alcohols, is 9 kg CO_{2eq}/L, which is significantly greater than that of wine. The difference is attributed mostly to a greater reliance on air freight for transportation to market and secondarily to distillation [25]. The preceding table gives the breakdown by inputs into the process.

12.2 Automobile industry

This section is devoted to the manufacturing and recycling of vehicles. Use of vehicles is treated in Chapter 3, and transportation aspects in Chapter 5. As a major industrial sector, automobile manufacturing and its upstream sourcing of materials cause significant impacts on the environment. For example, the production of 15 million light vehicles (about the annual output of North America) requires almost 30 million tons of steel. Significant fractions of the plastics (7%) and aluminum (5%) markets are devoted to the automobile industry ([26] page 11). It was estimated that, in 2001, the 217 million vehicles on US roads contained 5.3% of all the steel and 13.8% of all the aluminum in use in the country at the time ([27] page 1).

12.2.1 Manufacturing

The following table lists the number of light vehicles (passenger cars and SUVs) manufactured and used per population in various regions of the world.

Light vehicles produced and used across the world		
Region	2019 Production (vehicles/year)	2015 Motorization (vehicles in use/1,000 people)
Africa	1,105,147	42
North America	16,783,398	670
United States	10,880,019	821
Central and South America	3,319,361	176
Asia, Middle East, and Oceania	49,266,873	105
Australia	5,606	718
China	25,720,665	118
India	4,516,017	22
Japan	9,684,298	609
Philippines	8,400	38
South Korea	3,950,617	417
Europe	21,312,082	471
EU + EFTA[1]	17,735,151	581
Others	3,576,931	281

Continued

| Russia | 1,719,784 | 358 |
| World | 91,786,861 | 182 |

[1]EFTA = European Free Trade Association (incl. Norway, Switzerland)
Sources: [28,29].

The following table gives the typical material distribution in an automobile.

Material composition of a generic passenger car					
Metals	Ferrous	985 kg		64.3%	
	Aluminum	96.6 kg		6.3%	
	Copper	18 kg	1,123 kg	1.1%	73.3%
	Lead	13 kg		0.85%	
	Other	9.8 kg		0.64%	
Plastics	Polyurethane	35 kg		2.3%	
	PVC	20 kg	143 kg	1.3%	9.3%
	Other	88 kg		5.7 %	
Rubber		45 kg		2.9%	
Glass		42 kg		2.7%	
Fluids		74 kg		4.8%	
Misc.		105 kg		6.9%	
Total		1,532 kg		100%	

Source: [26] Appendix A, Table A1.

The following table lists the environmental impacts in the production of a typical nonelectric car.

Impacts in the production of a non-electric car					
Life-cycle stage	Energy consumption	Water used	CO_{2eq} emissions	VOC emissions	Solid waste
Total production	8,300 MJ/car	3.4 m^3/car	530 kg/car	2.25 kg/car	75 kg/car
Material acquisition	82%	91%	79%	—	—
Manufacturing and assembly	18%	9%	21%	—	—

Sources: [30] for absolute numbers, [26] for proportions.

From the preceding table, we see that during the production of a nonelectric car, there are more greenhouse gases emitted (530 kg) than solid waste (75 kg). As the average mass of a generic nonelectric vehicle is 1,532 kg (see above), the amount of greenhouse gas emissions in its production amounts to about 35% of its mass, while the amount of solid waste amounts to only 5%.

The impact profile of an electric car is quite different from that of a conventional car with an internal combustion engine (ICE). Although the electric car has a smaller footprint during its use (not the topic here), its production footprint is significantly greater than that of a conventional car of the same size.

For the average European Union electric car ([31] Fig. 1), the carbon footprint of battery manufacturing is estimated at 36.1 g CO_{2eq}/km for a 150,000 km life use, that is, 5,410 kg CO_{2eq}. The rest of the manufacturing amounts to 5,570 kg CO_{2eq} (37.1 g/km). This makes the total production footprint be 10,980 kg CO_{2eq}, which far exceeds that of the ICE car at 530 kg CO_{2eq} (preceding table). It also exceeds the electric car's footprint during the use of 57 g CO_{2eq}/km (depending on the local electricity mix) for 150,000 km, or 8,550 kg CO_{2eq}. [*Note*: Beware of older studies. Older electric cars had significantly smaller battery packs in return for a much shorter range.]

12.2.2 End of life

A vehicle reaches the end of its life around 10−15 years ([27] page 3), after it has been driven for an average of 150,000 km (=93,000 mi) ([32] page 56).

Because it is typically a low-budget business, the dismantling, shredding, and recycling sector of the automobile industry is very sensitive to the fluctuating prices of commodities and is a sector in constant flux. Consequently, available numbers may not be very reliable. The one element that keeps the sector operating is the presence in many countries of regulations that require the capture and partial recycling of end-of-life vehicles.

It is estimated that over 90% of obsolete vehicles are collected and dismantled to some extent. Waste is reduced by the capture of reusable parts and recycling of a variety of materials, with an estimated 80% of the weight avoiding the landfill. Of this, 65−70% consist of metallic components with the rest (10−15%) consisting of plastics, glass, rubber, and other materials that require a higher degree of dismantling before they can be recycled ([26] page 23). Technically, up to 90% of most car bodies are recyclable ([27] page 43), but this level is not achieved for economic reasons.

The following table illustrates the types and amounts of materials that are recycled, partially recycled, or landfilled.

The degree to which materials are recycled from a light vehicle					
Mostly recycled		Partially recycled		Mostly landfilled	
Material	% mass of whole vehicle	Material	% mass of whole vehicle	Material	% mass of whole vehicle
Steel	55.6%	Plastics[1]	8.1%	Rubber	4.6%
Iron castings	7.9%	Fluids and lubricants	5.27%	Glass	2.6%
Copper and brass	1.3%	Lead	1.3%	Textiles	1.1%
Aluminum	7.7%	Zinc castings	1.03%	Coatings	0.7%
		Magnesium castings	0.25%		
		Powder metal parts	0.22%	Other	2.3%
		Misc. metals	1.05%		
Total	72.5%	Total	17.2%	Total	11.2%

[1]Almost exclusively PP and ABS plastics ([27] page 38).
Source: [27] Figs. 1.7 to 1.9.

It is worth noting that cars have been made lighter to improve fuel efficiency, and this has resulted in a greater use of fiber composites or other complex materials in return for lesser fractions of recyclable metals. The consequence is that cars with high fuel efficiency (including hybrids) have lower recyclability rates (around 74%) than more conventional cars (upward of 80%) ([27] pages 13−14).

Automotive shredder residue contains 40−55% of hydrocarbon-based materials (mostly plastics, also some oils, fibers, and wood), which can be incinerated. Incineration of shredder residue yields 5,400 BTUs/lb (=13 MJ/kg) in average ([27] page 18). On-site electricity generation exceeds the need of the shredder and can be sold ([27] page 64).

The net energy consumption in end-of-life processes, after discounts for energy recovery, is estimated at 2 GJ for a typical mid-size sedan [33].

12.3 Building construction

Numbers in this section relate to the construction of buildings. See Chapter 6 for numbers pertaining to the use of buildings once they are constructed.

At the global level, civil works and building construction consume 60% of raw materials. Of this fraction, building construction accounts for 40%. In Europe, mineral extraction for building construction amounts to 4.8 metric tons per person per year, which is about 64 times the person's weight. In Spain, every habitable square meter of a typical building requires a total of 2.3 metric tons of more than 100 different materials, not counting materials wasted along the way [34].

The following table provides the energy, carbon, and water footprints of materials commonly used in building construction.

Energy, carbon, and water footprints of building materials				
Category	Material	Primary energy (MJ/kg)	Global warming potential (kg CO_{2eq}/kg)	Water demand (L/kg)
Stone-like	Cement	4.235	0.819	3.937
	Concrete	1.105	0.137	2.045
	Reinforced concrete	1.802	0.179	2.768
	Brick	3.562	0.271	1.890
	Ceramic tile	15.65	0.857	14.45
	Fiber cement roof slate	11.54	1.392	20.37
Metals[1]	Aluminum	136.80	8.571	214.3
	Copper	35.59	1.999	77.79
	Reinforced steel	24.34	1.526	26.15
Wood products	Softwood timber (sawn, planed, and kiln dried)	21.00	0.3	5.119
	Laminated timber	27.31	0.541	8.366
	Particle board	34.65	0.035	8.788
	Oriented strand board (OSB)	36.33	0.62	24.76

Continued

	Cellulose fiber	10.49	1.831	20.79
	Cork slab	51.52	0.807	30.34
Insulation	Polyurethane rigid foam	103.8	6.788	351.0
	Rock wool	26.39	1.511	32.38
	Glass	15.51	1.136	16.54
Miscellaneous	PVC	73.21	4.267	512.0
	Roof tile	4.590	0.406	2.456

[1]*Numbers here are for virgin materials. The use of recycled steel and aluminum can yield savings of more than 50% in embodied energy in the building ([34], page 1135).*
Source: [34] Tables 2 to 6.

Typically, structural materials account for more than 50% of the embodied energy in the building. Savings amounting to 50% over the 50-year lifetime of the building can be obtained by the use of alternative materials, such as hollow concrete blocks, stabilized soil blocks or fly-ashes, instead of materials with a high embodied energy such as reinforced concrete ([34] page 1135).

The energy and water embodied in buildings per unit floor area are 25.8 GJ/m^2 and 54.1 m^3/m^2, respectively ([35] Tables 1 and 2). Including only the impact associated with materials, every m^2 of floor area entails on average an energy consumption of 5,754 MJ and an emission of 500 kg of CO_{2eq} (both of which are variable depending on the type of building) ([34] p. 1134). Using these numbers, the primary energy needs and greenhouse gas emissions for the production of materials needed in European building construction can be estimated per m^2 of gross floor area as done in the following.

Footprints of building materials per m^2 of floor area				
Material	Primary energy (MJ/m^2, %)		CO_{2eq} emission (kg CO_{2eq}/m^2, %)		Water (m^3/m^2)
Steel	1,467	25.5%	93.5	18.7%	9.4
Ceramic	1,237	21.5%	101.5	20.3%	–
Cement	673	11.7%	151.5	30.3%	3.6
Mortar	523	9.1%	34.5	6.9%	–
Aluminum	443	7.7%	11.5	2.3%	–
Additives	230	4.0%	7.5	1.5%	–
Gravel	201	3.5%	14.5	2.9%	–
Lime	173	3.0%	39.5	7.9%	–

Continued

Prefabricated concrete	115	2%	10.0	2%	—
PVC	109	1.9%	5.0	1.0%	—
Wood	86	1.5%	5.5	1.1%	—
Other	506	8.8%	25.0	5.0%	41.1
Total	5,754	100%	500	100%	54.1

Note: Some round-off errors.
Sources: [34] Figs. 1 and 2 with percentages converted into absolute values, [35] Fig. 5.

12.4 Chemical industry

For feedstock and energy demand incurred in the production of specific chemicals, see Section 1.4.

The chemical industry, including the petrochemical industry (oil refineries), is energy-intensive, accounting for some 47% of total energy use by all industries combined ([36] pie chart), but it is comparatively low in carbon emissions, accounting for only 8% of the carbon emissions of all industries combined [37] and very water-intensive, accounting for 43% of all industrial water uses ([38] Fig. 14). In the United Kingdom, energy consumption by the chemical energy (about 220 PJ/yr) is highest of all sectors, well above the food-and-drink industry (about 150 PJ/yr) and has the second-largest carbon footprint (19% of all greenhouse gas emissions) after the steel industry (25%) [1].

Pollution intensity of the chemical industry varies across sectors as the following table reveals when impacts are expressed in kg of emission per $1 million of production output. Although these numbers are from circa 2000 and the value of the dollar has changed over time, the relative comparisons across columns and rows should remain indicative.

Emissions from the chemical industry by sector and type of pollutant				
	Pollutant (in kg)	Industrial chemicals	Fertilizers and pesticides	Synthetic resins, fibers and plastic materials
Air	Sulfur dioxide (SO$_2$)	2,179	42	672
	Nitrogen dioxide (NO$_2$)	1,779	443	690
	Carbon monoxide (CO)	217	2	64
	Volatile organic compounds (VOCs)	354	23	1,487

Continued

	Total particulates (TP)	517	114	197
	Toxic chemicals	369	110	628
	Bioaccumulative metals	7	0	1
Water	Biochemical oxygen demand (BOD)	15	4	34
	Suspended solids (SS)	201	94	79
Land	Toxic chemicals	410	63	239
	Bioaccumulative metals	112	0	40

Source: [38] Table 4.

Bayer AG mentions in its 2019 Annual Report ([39] page 58) that its energy footprint 247 kWh/€ (794 MJ/$), which, given the wide variety of chemicals produced by this company, may be representative of the chemical industry as a whole.

By contrast, the One Linde Group (incl. the acquisition of Praxair), which specializes in the production of gases, reported to have consumed 135.7 million GJ of energy and emitted 24.9 million metric ton of CO_{2eq} in 2018 for sales of $28 billion [40]. This amounts to 4.85 MJ/$ and 0.89 kg CO_{2eq}/$, a much leaner operation.

Specific information on the oil and gas industry can be found in Section 12.9.

12.4.1 Toxic release inventories

In the United States, all industrial facilities are required to report annually how much of each chemical from a list they have released that year in air, water, or land. The outcome is the Toxic Release Inventory, which serves as a repository of information on the chemical industry. The obligation to report also prompts polluting companies to treat or recycle substances instead of releasing them.

The US Environmental Protection Agency [41] estimates the various fates of toxic substances as tabulated in the following.

Fate of toxic substances produced in the United States	
Fate	Fraction
Recycling	53%
Treatment	26%

Continued

Energy recovery (incineration)	10%		
Disposal or other releases	12%	Disposal on land	8.16%
		Release to the air	1.92%
		Release to the water	0.60%
		Off-site disposal and other forms of release	1.32%
Source: [41].			

The breakdown of production and disposal of toxic substances is as tabulated in the following. The chemical industry is by far the dominant producer of toxic substances and the second-highest in disposing of them (after the metal mining industry). In 2018, the US chemical industry was able to recycle, treat, or incinerate 97% of its unwanted toxic substances, releasing only 3% to the environment ([42] page 13).

Toxic substances produced and disposed in the United States during 2018		
Industrial sector	**Fraction produced**	**Fraction disposed**
Chemical manufacturing	55%	14%
Primary metals	8%	9%
Petroleum product manufacturing	7%	(with "other")
Metal mining	6%	47%
Food processing	5%	4%
Paper manufacturing	4%	4%
Electric utilities	4%	9%
Hazardous waste management	(with "other")	4%
Other sectors not listed separately	11%	9%
Total	100% 32.12×10^9 lbs	100% 3.80×10^9 lbs
Source: [42].		

12.4.2 Accidents

History shows that the chemical industry is prone to cause occasional but highly visible and fatal accidents. The following table recalls accidents over the last 50 years with at least 20 casualties.

Fatal accidents caused by chemicals				
Date	**Location**	**Company/site**	**Chemical**	**Fatalities**
Feb. 1971	Woodbine, Georgia	Thiokol chemical plant	Explosives	24
Jun. 1974	Flixborough, UK	Nypro UK	Cyclohexane	28
Jul. 1976	Seveso, Italy	ICMESA plant	Dioxins	3,000 animals
Feb. 1984	Cubatao, Brazil	favela of Vila Sao Jose	Gasoline	>100
Dec. 1984	Bhopal, India	Union Carbide	Methyl isocyanate	>3,700
Oct. 1989	Houston, Texas	Phillips Petroleum Co.	Flammable gases	23
Sep. 2001	Toulouse, France	AZF Fertilizer Co.	Ammonium nitrate	29
Feb. 2008	Istanbul, Turkey	Fireworks company	Explosives	22
Aug. 2014	Kaohsiung, Taiwan	Gas distribution	Natural gas	32
Aug. 2015	Tianjin, China	Port of Tianjin	Ammonium nitrate	173
Aug. 2020	Beirut, Lebanon	Port warehouse	Ammonium nitrate	181
Source: [43].				

12.5 Consumer products

12.5.1 Furniture

As a rule, the carbon stored in wood products, such as those used in construction, compensates for the CO_2 emissions caused during their production [44]. For furniture, there is a net positive carbon footprint due to finishing varnishes and upholstery. Also, some items, like swivel desk chairs, may not contain any wood.

For the average UK household, home furnishings are responsible for a carbon emission of around 1.6 metric tons CO_{2eq} or 6% of the household total carbon footprint [45].

The following table provides the estimated net carbon footprint of various furniture items.

Carbon footprint of furniture		
Category	**Item**	**Average carbon footprint (kg CO_{2eq})**
Bedding	Sofa bed	88
	Double mattress	79
	Double bed (divan)	35
	Single bed (divan)	33
	Headboard	22
Dining	Dining chair	27
	Dining table	25
Kitchen	Storage unit (1 m)	42
	Storage unit (0.5 m)	40
	Drawer line unit (1 m)	40
	Appliance housing	35
	Drawer line unit (0.5 m)	29
	Countertop	26
	Wall cabinet unit (1 m)	25
	Base sink unit	22
	Wall cabinet unit (0.5 m)	18
	Full height base unit	17
Office	6-person conference table	228
	Desk chair	72
	Tambour storage unit	50
	Wooden filing cabinet	48
	Work station (1.6 m × 1.2 m)	45
	Steel pedestal	44
	Visitor chair	36
	Rectangular desk (1.6 m × 1.8 m)	35
	Bookcase	18

Continued

Seating	Sofa	90
	Reclining chair (electric)	75
	Armchair	43
	Footstool	17
Miscellaneous	Waiting room bench seating	66

Source: [46].

12.5.2 Personal hygiene

The wastewater flowing from factories that manufacture liquid detergents, such as shampoos and soaps, carry a chemical oxygen demand (COD) in the range of 500–750 mg/L, a biochemical oxygen demand (BOD) in the range of 1,600–2,200 mg/L, and total suspended solids in the range of 110–140 mg/L [47].

The approximately 3,800 disposable diapers that a child needs for the first two and a half years of life necessitate a total of 898 L (=237 gallons) of crude oil for their plastic waterproof lining, 324 kg (=715 lbs) of plastic, and pulp from 4.5 trees for the fluffy padding. The reusable diapers would have needed 85,000 L (=22,500 gallons) of water to clean them, enough drinking water to quench the thirst of a person for 93 years [48].

12.5.3 Beverages

The beverage industry uses large quantities of water. The ratio of water volume used in the plant to the water volume in the beverages varies from 2.9 to 4, with the excess water exiting the plant as wastewater [49]. If reusable bottles are used, their washing can account for as much as 60% of the water used. Bottle-washers need 150–200 mL of water per bottle [49].

The water footprint varies with the type of beverage with milk and apple juice being particularly consumptive, as indicated in the following table.

Water footprint of selected beverages		
Beverage	**Beverage quantity**	**Water need**
Coca Cola	1 L	1.92 L
Milk	200 mL	200 L
Beer	250 mL	75 L
Tea	250 mL	35 L
Wine	125 mL	120 L
Apple juice	200 mL	190 L
Orange juice	200 mL	170 L

Sources: [50],[51] Table 4.2.

12.5.4 Tobacco

Globally, 3.37 million hectares were dedicated to harvesting tobacco in 2018, representing a slow decrease over the decades [52], but still large swaths of forests and woodlands are cleared every year for tobacco farming (200,000 hectares/year in the 1990s, resulting in 1.7% of global net losses of forest cover) [53].

The production of 1 million cigarettes[1] (about 10^3 kg) requires 0.88 ha (2.2 acres) of land to grow 5,400 kg of green tobacco leaves yielding 1,080 kg of dry tobacco, 3,700 m^3 of water, and 10.4 GJ of energy, and releases 4,200 kg of solid waste, 9,200 L of wastewater, and 14,000 kg CO_{2eq} ([55] Fig. 1).

The packaging of 1 million cigarettes (20 cigarettes per pack, each empty pack weighing about 6 g) demands 300 kg of paper, foil, cellophane, glue, and ink [56], most of which ends up in landfills.

The combustion of 1 million cigarettes emits 430 kg of CO_2 and 870 kg of methane, for a total global warming potential of 28,000 kg CO_{2eq}. It is also responsible for the discarding of 170 kg of nonbiodegradable filter waste ("cigarette butts"), most of which is ending up as litter [56].

12.6 Energy industry

Note: This section is devoted to the conversion from primary sources to utilizable forms of energy. For the utilization of energy itself, the reader is referred to Chapter 3.

The energy supply sector suffers major inefficiencies in converting and transporting energy. In the United States during 2018, 25.3 quads of energy were lost in the process of delivering 75.95 quads of utilizable forms of energy [57], a ratio of 0.333 units of primary energy lost for each unit of utilizable energy delivered. In other words, the energy supply sector converts 75.0% of primary energy sources into usable forms such as electricity, heat, and refined oil products, while losing 25.0% along the way. A significant portion of the 25.0% loss is the low 39% overall efficiency in electricity generation [58]. For electricity transmission losses, see Section 3.6.1.

A consequence of fuel combustion during electricity generation and other conversion inefficiencies is that the energy supply sector is a major emitter of greenhouse gases. In the United States during 2018, 27% of total anthropogenic greenhouse gas emissions were attributed to electricity generation, which is almost the entirety of the emissions from the energy supply sector [59]. The result is that the carbon intensity of energy in the United States is 48 kg CO_2 per million BTUs of energy input [60]; this amounts to 0.045 g CO_2/kJ or, with a 39% conversion efficiency to electricity, 0.420 kg CO_2 per

---

1. The global production of cigarettes was 5.7 trillion in 2016 [54].

kWh of electricity generated (close to the value 0.45 kg CO_{2eq}/kWh quoted in Section 4.34 for the electricity mix in the United States).

Fossil fuel extraction and distribution contribute about 5—10% of the total fossil-fuel greenhouse gas emissions ([61] page 528). The extraction of oil sands and their refining into gasoline cause emissions estimated at 35—55 g CO_{2eq}/MJ of fuel, compared to emissions of 20 g CO_{2eq}/MJ for the production and refining of regular petroleum ([61] page 528).

The environmental impacts of electricity generation are tabulated in the following table by energy source. Hydropower and biofuels top the list for water consumption per unit of electricity, whereas solar and wind have insignificant water footprints.

		Environmental impacts of electricity by energy source			
		Greenhouse gas emissions (kg CO_{2eq}/MWh)			
Energy source	Total[1] water consumed (m^3/MWh)	Infrastructure and supply chain	Direct emissions from fuel combustion	Methane emissions along the way	Total lifecycle
Biofuel (ethanol)	53.1	45	0	0	45
Biodiesel	57.5	—	0	0	—
Coal	1.84	16.8	1,047	39.2	1,103
Geothermal	0.86	(no data)			
Hydropower	16.8	0	0	26	26
Natural gas	0.98	61	426	95	582
Nuclear	2.00	29	0	0	29
Oil	1.56	256	—	—	733
Solar (PV)	0.02	85	0	0	85
Wind	0.001	26	0	0	26
Weighted average	2.18	(no data)			

[1]*For both fuel processing and power plant operations, excluding water withdrawal for cooling.*
Sources: [62] Table 1, [63] Table 1, [64—66], [67] Fig. ES-2 and Exhibit A-7.

The International Hydropower Association published a graph ([68] Fig. 2) that shows a strong inverse relationship between greenhouse gas emissions E (in kg CO_{2eq}/MWh) and the electric power P generated per area of reservoir behind the dam (in W/m^2):

$$E = \frac{90}{P}.$$

The emission value $E = 26$ kg CO_{2eq}/MWh quoted in the preceding table corresponds to a power density $P = 3.5$ W/m^2, which is an average value among hundreds of the hydropower systems across the world ([68] Fig. 2). Note that the constant value 90 in the numerator of the equation corresponds to a constant rate of greenhouse gas emissions (mostly methane) per area independent of the power generated and equal to:

$$E \times P = 0.090 \frac{\text{g } CO_{2eq}}{\text{m}^2 \times \text{hour}} = 0.79 \frac{\text{kg } CO_{2eq}}{\text{m}^2 \times \text{year}}.$$

A separate study notes the same inverse relation and quotes the following emission rates as a function of the climate zone.

Carbon dioxide and methane emissions from reservoirs		
Climate zone	kg of CO_2/(m^2.year)	g of CH_4/(m^2.year)
Boreal	0.97	40
Temperate	0.42	7.2
Tropical	1.2	46

Source: [69] Table 2.

12.7 Healthcare

12.7.1 Healthcare facilities

Annually, an average inpatient healthcare facility consumes 240,000 BTUs per ft^2 (=70 kWh/ft^2 per year =86 W/m^2) in total for all forms of energy, including heating, cooling, and ventilation ([70] page iv; [71] Fig. 4). The annual electricity consumption of an in-patient health care facility amounts to 31 kWh/ft^2 (25.8 kWh/ft^2 for all healthcare facilities combined as quoted in Section 6.3.3), second highest among commercial building types after food sale facilities (48.7 kWh/ft^2) and about twice as high as office buildings (15.9 kWh/ft^2) ([72]).

The water consumption in hospitals (inpatient healthcare facilities) is estimated at 49.3 gallons per ft^2 per year (5.5 L/m^2/day) ([73] Fig. 1). Water consumption per bed and per employee is given in Section 2.4.

The majority (approximately 85%) of solid waste from healthcare facilities is similar to commercial solid waste and includes office paper, cardboard, plastics, metals, glass, and food wastes. The remaining 15% is medical waste handled as potentially infectious, often referred to as "red bag" or "regulated medical waste" ([70] page v).

Regulated medical waste, which is potentially infectious, is often incinerated on-site, leading to problematic air emissions. It is estimated that fumes from medical waste incinerators are some of the largest sources of atmospheric emissions of mercury and dioxins ([70] page 11). The following table lists the sources and quantities of regulated medical waste in the United States.

Sources of regulated medical waste in the United States		
Facility	lbs/month	kg/year
Hospitals	8,400	46,000
Laboratories	600	3,300
Clinics	180	980
Physicians' offices	24	130
Dentists' offices	13	71
Veterinary facilities	20	110
Long-term healthcare facilities	390	2,100
Blood banks	440	2,400
Funeral homes	32	170
Source: [74] Table 4.		

The emission factors for controlled air medical waste incinerators are tabulated in the following.

Emission factors for medical waste incineration	
Pollutant	Emission (g/kg of waste)
Carbon monoxide (CO)	1.48
Dioxins (total CDD)	1.07×10^{-5}
Lead (Pb)	3.64×10^{-2}
Mercury (Hg)	5.37×10^{-2}
Nitrogen oxides (NO_x)	1.78
Particulate matter (PM)	2.33
Sulfur dioxide (SO_2)	1.09
Total organic compounds	0.150
Source: [75] Tables 2.3-1, -2, -7 and -11.	

Because most of the radioactive isotopes used in medicine are short-lived, their treatment simply consists of letting them age in the facility until they are decayed to safe levels. The remaining part, consisting of long-lived radioactive medical waste, is disposed of in specialized facilities for low-level radioactive waste, to which it contributes about 1% ([70] page v).

Laundry volume per patient per day exceeds 20 lbs (=9 kg) ([70] page 6). For the environmental impacts of laundry, See end of Section 9.5.

The UK National Health Service publishes annually a breakdown of its GHG emissions [76]. The following table provides the numbers reported in 2016. See the full report for similar breakdowns of water usage, air emissions, and solid waste.

Relative carbon footprints in the healthcare sector		
Scope	Sub-area	Percentage of total GHG emissions
Core operations	Fuel use	8.6%
	Electricity	7.6%
	Business travel and fleet	4.6%
	Anesthetic gases	1.7%
	Water	1.5%
	Waste	0.1%
	Subtotal	24.2%
Supply chain	Instruments and equipment	13.2%
	Pharmaceuticals	12.1%
	Business services	9.2%
	Food and catering	5.9%
	Chemicals, gases, and fuels	4.4%
	Freight transport	3.2%
	Construction	3.0%
	Paper products	2.8%
	Other products	2.3%
	ICT	1.3%
	Subtotal	57.4%

Continued

Community	Patient and visitor travel	7.0%	
	Staff commuting	3.9%	
	MDI usage	3.1%	
	Subtotal		14.0%
Outsourced services	Subtotal		4.4%
	Total		100%

Source: [76] 2018 Report based on 2017 data.

12.7.2 Pharmaceuticals

The pharmaceutical industry has a fundamental issue, namely, it takes large quantities of energy and feedstock materials to produce small quantities of specialty chemicals. The production of Paramax tablets is a good example, as shown in the following table. In other words, the pharmaceutical industry produces far more by-products than products.

Material attrition in the making of a drug			
Stage	Product	Amount (metric tons/ year)	Mass loss from previous stage
Basic feedstock	Propylene	12 million	—
Basic building blocks	Acrylonitrile	4.3 million	64%
	Acrylamide	300,000	93%
Intermediates	2-Diethylamino ethylamine	200	99.9%
Bulk activities	Metoclopramide	50	75%
End product	Paramax tablet	10^9 tablets with 5 mg of metoclopramide	90%

Source: [77] Fig. 2.1.

The breakdown of materials in the production of a drug is: 56% solvents, 32% water 7% reactants, and 5% other [78].

While the average commercial building consumes around 81 kBTUs/ft^2 (see Section 6.3.2), the average pharmaceutical plant consumes 1,210 kBTUs/ft^2 (=3,820 kWh/m^2), 65% of which is attributed to heating, cooling, and ventilation to maintain strict ambient air conditions [79].

The following table provides the breakdown of energy use in the pharmaceutical industry.

Energy use in the pharmaceutical industry					
Bulk manufacturing	Research and development	Formulation, packaging, and filing	Offices	Warehouses	Misc.
35%	30%	15%	10%	5%	5%
Source: [80].					

The carbon intensity of the pharmaceutical industry (averaged over the 15 largest companies) was 48.6 metric tons of CO_{2eq} per $1 million of revenue in 2015 and projected to decrease to about 31.2 t CO_{2eq}/$M by 2020 ([81] Fig. 4). There is a spread factor of 5.5 between the most carbon-intensive (77.3 t CO_{2eq}/$M) and the least carbon-intensive (14 t CO_{2eq}/$M) ([81] p.193).

12.8 Media and entertainment

This section concerns the impacts within the media and entertainment sectors. For the consumption of media contents, see Section 8.3.

12.8.1 Printing industry

The impacts of the printing industry are dominated by those of the paper it prints upon and secondarily by transportation. The actual printing is a minor contributor. For example, the CO_2 emissions of printing one page on a laser printer, which is the most energy-intensive printer, is 1.03 g [82] whereas the carbon footprint of the sheet of paper is 5.23 g (Section 1.3). With an average of 1.65 g CO_2/sheet or 0.82 g CO_2/page (2 printed pages per sheet), offset industrial printers have similar carbon footprints, as the following table indicates.

Environmental impacts of printing

Indicator	Coldset web offset	Heatset web offset	Sheetfed offset	Gravure	Electro-photography	High speed inkjet	Units[1]
Paper use	1.06–1.1	1.2–1.24	1.1–1.3	1.09–1.14	1.05–1.17	1.06–1.11	t/t
Ink use	14–20	26–32	5–8	22–50	11–39	10–40	kg/t
Energy	420–500	750–850	800–1,040	590–650	300–700	450–800	kWh/t
CO_{2eq}[2]	210–250	375–425	400–520	295–325	150–350	225–400	kg/t
VOCs	0.1–0.3	0.5–0.6	3.0–5.3	1.2–1.9	0	0	kg/t
Hazardous waste	2.7–3.0	1.8–2.0	1–6	0.98–1.1	0.4–1.3	0	kg/t

[1] Numbers are for one metric ton of printed product. One metric ton of paper contains 440 reams, with each ream containing 500 sheets, for a total of 220,000 sheets of paper at 4.55 g per sheet.
[2] Carbon emissions are calculated from energy with electricity footprint of 0.50 kg CO_{2eq}/kWh.
Source: [83] Table 4.

The carbon footprint of a 0.5 kg book breaks down as tabulated in the following.

Carbon footprint of a 0.5 kg book	
Life cycle stage	**g CO_{2eq}/book**
Chemical, materials, and fuels	40
Inner sheets	500
Cover	70
Endpapers	10
Direct emissions during printing	0
Energy purchased for the printing	500
Transportation	40
Total	1,160

Source: [83] Table E5.1.

The carbon footprint of the printing press varies widely with the equipment. One mention is that 260 metric tons of CO_2 are generated during the manufacturing of a Speedmaster XL 106-6+L (capable of printing 18,000 sheets/hour) [84].

12.8.2 Film and television industry

The film and television industry in California (concentrated in the Los Angeles area) consumes about 105×10^{15} J/year and emits roughly 8.4 metric tons of CO_{2eq}/year. This is very similar to the hotel sector (110×10^{15} J/year, 9×10^6 CO_{2eq}/year) and the apparel sector (125×10^{15} J/year, 10×10^6 CO_{2eq}/year) and about half of the semiconductor industry (205×10^{15} J/year, 15×10^6 CO_{2eq}/year) ([85] Figs. 2 and 3). It also ranks relatively high in terms of hazardous waste generated (200,000 metric tons/year) and number of fatalities (about 18 deaths/year) ([85] Figs. 4 and 5).

The television industry generates 13.5 metric tons of CO_{2eq} for every hour of on-screen content [86]. The breakdown of carbon emissions is tabulated in the following table. Personnel travel including hotel accommodations accounts for about half of the carbon footprint.

Carbon footprint of the television industry by activity		
Activity	**Fraction**	**kg CO_{2eq} per hour of show**
Material	0.7%	100
Production office	17.6%	2,370

Continued

Personnel travel	33.9%	4,580
Accommodation (hotels)	17.0%	2,290
Studios	6.2%	840
On location	10.4%	1,400
Equipment transport	11.6%	1,560
Post production	2.7%	360
Total	100%	13,500

Source: [86].

The carbon footprint can also be estimated by genre, as tabulated in the following.

Carbon footprint of television programs by genre	
Genre	Metric tons of CO_{2eq} per hour of viewing on television screen
Children's programs	7.8
Comedy	18.4
Current affairs (news)	7.3
Drama	44.1
Entertainment	10.5
Factual	13.1
Sports	3.3
Average	13.5

Source: [86] page 7.

12.8.3 Music industry

The energy and carbon emissions associated with delivering one album of music to the customer varies from a high of 53 MJ/album and 3,200 g CO_2/album for a compact disk (CD) purchased via traditional retail, down to 7 MJ/album and 400 g CO_2/album for a downloaded file with no subsequent CD burning [87].

The following is the breakdown of greenhouse gas emissions in the UK music industry.

Carbon footprint across the UK music industry	
Activity	**Fraction**
Audience travel	43%
CD life-cycle emissions	26%
Concert venues	23%
Electric generators	4%
Equipment transport by truck	2%
Tour buses	1%
Offices	1%
Source: [88] Fig. 2.	

The life-cycle analysis of the CD album supply chain leads to the following greenhouse gas emissions. The top two contributors are the plastic case with inside tray (accounting for a third of the entire carbon footprint!) and in-store retail.

Carbon footprint of CD albums by life stages			
Stage of CD lifecycle		**g of CO_{2eq}/ CD**	**Fraction**
Recording studio	Energy emissions	37	3%
	Waste emissions	3	<1%
Manufacturing	Energy emissions	100	9%
	Materials life cycle emissions (excl. packaging)	100	9%
	Plastic jewel case and tray	376	34%
	Paper booklet and insert card	64	6%
	Waste emissions	6	1%
Marketing	Office energy emissions	114	10%
Transport	From factory to warehouse	76	7%
	From warehouse onwards	51	5%
	Transit packaging	7	1%
	Waste emissions	28	3%
Retail	Retail store energy emissions	132	12%

Continued

Record label and publishing	Office energy emissions	2	<1%
	Office waste emissions	1	<1%
Promotion	Business travel—hotel	7	1%
	Business travel—train and taxi	2	<1%
	Business travel—airplane	5	<1%
Total		1,111	100%

Source: [88] Table 2.

12.9 Oil and gas industry (Petrochemical industry)

Oil refineries are among the most energy and carbon-intensive sectors of the chemical industry. The Union of Concerned Scientists [89] reports that a typical oil refinery in the United States consumes between 540 and 690 MJ and emits between 97 and 111 lbs of CO_2 per barrel[2] of crude oil refined. Globally, the carbon intensity of oil refining is estimated at 10.3 g CO_{2eq} per MJ of crude oil processed (within a range of 3.3—20.3 g CO_{2eq}/MJ) [90].

Energy inputs have been tallied for the extraction of oil and gas, including support, drilling, and extraction. The numbers tabulated in the following table are in J per 100 J of crude oil output.

Energy inputs for the extraction of oil and gas			
Input type	Support for oil and gas operations[1]	Drilling of oil and gas wells	Extraction of oil and gas
Diesel and distillate fuel oils	0.30 J	0.70 J	0.42 J
Residual fuel oil	0.23 J	0.44 J	0.15 J

Continued

---

2. A barrel of oil holds 42 US gallons ($= 0.159$ m^3), weighs 136 kg and has an energy content of 5.80×10^6 BTUs ($= 6.12$ GJ).

Gas (natural or manufactured)	0.05 J	0.02 J	32.42 J
Gasoline	0.16 J	0.11 J	0.43 J
Electricity	0.16 J	0.03 J	4.00 J
Other energy inputs	0.98	0.43	0.00
Percentage energy of production of oil and gas	0.19	0.17	3.74

[1]Support activities comprise establishments primarily engaged in performing support activities for oil and gas operations (except site preparation and construction activities).
Source: [91] Table 10.

The following are inputs and emissions incurred during crude oil extraction and transportation.

Inputs and emissions during crude oil extraction and transportation			
	Onshore extraction	Offshore extraction	Transportation[1]
Raw material consumption (g per kg of crude oil produced)			
Coal	9.68	0.00	3.654
Oil	695.51	215.00	2.149
Natural gas	4.36	3.07	0.568
Limestone	1.84	0.00	0.697
Water (L per kg crude oil)	0.00030	0.00	0.00039
Air emissions (g per kg of crude oil produced)			
Benzene	0.00017	3×10^{-5}	4.54×10^{-7}
Methane (CH_4)	0.20565	0.01859	0.0291
Carbon monoxide (CO)	0.02833	0.00883	0.0117
CO_2 for fossil fuel sources	38.93	8.988	17.89
Formaldehyde	0.00	0.00303	6.08×10^{-6}
Non-methane hydrocarbons	0.11152	0.01510	0.1127
Nitrogen oxides	0.09849	0.00977	0.0728
Particulates (all)	0.13619	0.00017	0.0560
Sulfur oxides	0.26233	0.00093	0.122

Continued

Solid waste (g per kg of crude oil produced)			
Nonhazardous solid waste	0.00355	0.00	0.00135

Energy (MJ per kg of crude oil produced)			
Total primary energy	30.01	9.175	0.271
Fossil energy	29.99	9.175	0.266
Fuel energy	0.654	0.667	42.54

[1]Based on US domestic data
Source: [92] Tables 31 and 35.

12.9.1 Onshore extraction

For every kilogram of crude oil produced onshore, 0.7 L of wastewater is released to the surrounding environment, including 1.0×10^{-4} kg of oil and grease. The emission factor for the production of solid waste is estimated at 0.0098 g of solid waste for every kg of crude oil. Also, an estimated 0.47 kg of associated natural gas is produced for every kg of crude oil extracted from offshore wells ([92] page 57).

The following table lists the typical constituents and concentrations found in wastewater from onshore oil production.

Constituent	Median concentration (mg/L)
Arsenic	0.02
Benzene	0.47
Boron	9.9
Chloride	7,300
Mobile ions	23,000
Sodium	9,400

Source: [92] Table 19.

12.9.2 Offshore extraction

For every kilogram of crude oil produced offshore, 10.14 L of wastewater is released to the surrounding environment, including 2.8×10^{-4} kg of oil and grease. The emission factor for the production of solid waste is estimated at 0.0098 g of solid waste for every kg of crude oil. Also, an estimated 0.26 kg of associated natural gas is produced for every kg of crude oil extracted from offshore wells ([92] page 62).

12.9.3 Oil spills

Oil spills are caused by oil tanker accidents, ruptured pipelines, and mishaps onboard offshore oil platforms. They also differ by how much oil is spilled and which environment is affected. So, it is near impossible to draw general quantitative conclusions, and much of the evidence remains anecdotal.

The following table compares the consequences of some of the major oil spills that occurred at sea, from which lessons can be drawn about impacts.

Harm caused by some major oil spills at sea				
Date	March 1978	February 1983	March 1989	April 2010
Location	English Channel	Persian Gulf	Prince William Sound, Alaska	Gulf of Mexico
Ship/ platform	Amoco Cadiz	Nowruz Field Platform	Exxon Valdez	Deepwater Horizon
Oil spilled	67×10^6 gal	78×10^6 gal	10.8×10^6 gal	205.8×10^6 gal
	223,000 tons	260,000 tons	36,000 tons	686,000 tons
Oil recovered	(40% evaporated)	−	8%	25%
Shoreline affected	360 km	−	1,990 km	1,600 km
Birds harmed or killed	15,000 −20,000	−	100,000 −250,000	82,000
Other quantified biological damage	About 10,000 fish and 15.5 million bivalve killed	Over 500 sea turtles killed	2,302 seals killed	At least 6,400 marine mammals harmed or killed; at least 1,146 sea turtles killed, possibly as much as 6,165; 88,500 square miles of fisheries closed
Source of info	[93]	[94]	[95]	[96]

Other miscellaneous numbers: The rate of bird kill is about 0.1 bird per metric ton of crude oil [93]; trapped in sand, sediment-oil-agglomerates take at least 32 years to biodegrade and more than 100 years in the absence of contact with sediments [97]; shoreline habitats, including mussel beds, take about 30 years to recover [95]; bivalves and fish need 3—6 generations to recover (i.e., up to 30 years for those with a life expectancy of 5—10 years), and birds may need up to 60 years before populations reach their stable age distribution [93].

12.9.4 Refineries and distribution

An oil refinery is a petrochemical plant that separates ("refines") crude oil into its constituent groups. The relative amounts depend on the origin of the oil, for not all crude oils are the same. On average, crude oil yields the following outputs after refining.

Outputs of oil refining		
Refinery product	Output amount (kg of product/100 kg of crude oil)	
Gasoline (petrol)	42.9	
Low-sulfur distillate	13.0	
Jet fuel and kerosene	10.1	
High-sulfur-distillate	9.9	
Residual fuel oil	6.0	
Coke	5.9	
Special oil and lubes	4.7	
Waxes, asphalt, road oils	4.3	
Miscellaneous	0.4	
	Subtotal	97.2
Losses (incl. emissions)	2.8	
	Total	100.0

Source: [92] Table 37.

Among several other inputs (see Ref. [92] Table 39), an oil refinery uses water, at a rate of about 0.44 L of water per L of crude oil refined. For transportation fuel, the consumption is highest for gasoline (0.60—0.71 L water per L of gasoline produced) and lowest for jet fuel (0.09 L water per L of jet fuel produced) [98].

Along the way, the refining process causes a series of air emissions, as shown in the following.

Petroleum refining process emissions

Process	Units	Particulate	SO$_2$	CO	Non-methane hydrocarbons	NO$_2$	CO$_2$
Catalytic cracking	g/L of crackers feed	0.052	0.79	–	–	0.11	40.7
Fluid coking	g/L of cokers feed	1.5	–	–	–	–	–
Vapor recovery	g/L of refinery feed	–	0.077	0.012	0.002	0.054	–
Sulfur recovery	g/kg of sulfur removed	–	29	–	–	–	–

Source: [92] Table 41.

12.10 Paper industry

The consumption of paper per person increases with affluence in the country and decreases with its level of internet penetration [99]. The following table provides the consumption per capita for the main types of paper products across the world in 2008.

Consumption of paper products per capita in the world			
Type of product	Range	Mean	Units
Printing and writing	7.0−97.7	44.8	
Newsprint	4.1−85.0	23.0	
Household and sanitary[1]	0.7−26.3	11.4	kg per person per year
All types	31.4−332.6	148.6	

[1]*Paper towels, tissue paper, and toilet paper.*
Source: [99].

Paper is mostly (95%) produced from cellulosic fibers, either from trees or from recycled paper and cardboard. About 5% comes from nonwood sources such as bagasse, cereal straws, bamboo, and rice. Recycling is more intense in developed countries, and nonwood fibers are more common in developing countries [100].

Papermaking proceeds in two major stages, the production of pulp from fibers followed by drying on rolls to produce long sheets of paper (the "paper machine"). Water is needed to produce the pulp (fibers swimming in water), which is then evaporated during the drying process. Additional water is used during the bleaching process if it takes place. For every short ton of paper produced, 17,000 gallons of water is needed for chemical pulping while the kraft mechanical pulping may use as little as 4,500 gallons [101], most of which is treated and reused. An older report ([102] page 47) pegs the range between 4,000 and 12,000 gallons of water per short ton of pulp.

Beware of claims in the media that it takes up to 10 or even 20 L (2.6−5.2 gallons) of water to make a single A4-sheet of paper! Such alarming statements are usually based on a study ([103] page 12) that includes the amount of rain consumed by a tree during its growth, ignores recycling of paper, and also ignores the fact that most of the water in a paper mill is treated and reused. Using the higher number of 17,000 gallons of water per short ton quoted above and considering that one metric ton of paper makes 440 reams of 500 sheets, one obtains only 0.32 L per sheet. Actually, as office paper is generally made with the less-water-intensive kraft mechanical pulping to preserve long fibers, the better number to use is 4,500 gallons per short ton, which translates into 0.085 L of water per sheet of office paper.

For every short ton of paper produced, the US pulp and paper industry uses 0.65 tons of virgin wood pulp (dry weight) and 0.38 tons of pulp from recycled sources, with a 2.4% mass attrition rate [104]. The following table provides the breakdown of energy during the several stages of paper production. The numbers are per short ton of pulp.

Energy[1] use in the production of paper products		
Process stage	Energy use (10⁶ BTUs/ton of pulp)	
Wood preparation		
Debarking	0.10	0.45
Chipping and conveying	0.35	
Pulping		
Mechanical pulping	7.68	
Stone groundwood	5.11	
Refiner mechanical pulping	6.10	
Thermo-mechanical pulping	7.09	
Chemical-thermal-mechanical pulping	7.68	
Chemical pulping	2.68	
Kraft process	2.60	
Sulfite process	5.38	
Semichemical pulping	3.86	
Recycled paper pulping	1.30	
Kraft chemical recovery process		
Evaporation	3.86	
Recovery boiler	1.13	8.04
Recausticizing	1.02	
Calcining	2.03	
Pulp bleaching	2.3	

Continued

Paper and paperboard production	6.26	
Paper refining and screening	0.84	
Newsprint forming, pressing, and finishing	1.44	5.61
Newsprint drying	4.17	
Tissue forming, pressing, and finishing	1.82	9.77
Tissue paper drying	7.95	
Uncoated paper forming, pressing, and finishing	1.80	6.90
Uncoated paper drying	5.10	
Coated paper forming, pressing, and finishing	1.80	7.10
Coated paper drying	5.30	
Linerboard forming, pressing, and finishing	0.92	4.97
Linerboard drying	4.05	

[1]Energy units are 10^6 BTUs/million short tons of paper; the split is about 93% fuels and 7% electricity.
Source: [104] Table 2-2.

Drying pulp into paper requires large amounts of heat (usually in the form of steam produced from the combustion of on-site organic waste) to evaporate water. Thus, the papermaking process is the most energy-intensive of the several processes, accounting for about 45% of total energy use. Pulping is the next largest energy-consuming step. Mechanical pulping (extraction of fibers from lignin by mechanical means) consumes electricity to drive grinding equipment while chemical and semichemical pulping use a 75—25% mix of steam and electricity ([104] page 17).

12.10.1 Paper versus plastic

What is preferable at the grocery store: Paper or plastic bags? Although the answer to this question has been known for some time to be in favor of the plastic bag [105], it appears that the general public has not accepted it chiefly because paper comes from renewable trees and biodegrades in landfills whereas plastic comes from nonrenewable petroleum and does not decompose after disposal. The shortcoming of this conclusion is that it ignores other and crucial elements of the life cycle of both types of bags, chiefly the impacts during manufacture. It takes far more energy to turn a tree into paper than petroleum into a plastic polymer. Paper bags are also much heavier to transport.

The following table contrasts the properties and impacts of typical grocery bags.

| | Comparison of shopping bags | | | | |
	Bag type	Light plastic	Heavy plastic	Brown paper	Nylon-like reusable	Cotton reusable
	Material	HDPE	LDPE	Kraft virgin paper[1]	nonwoven PP	cotton
Single bag	Volume (L)	19.8	21.5	20.1	19.8	25.2
	Weight (g)	10	35	55.2	116	154
	Energy (kJ)	22.14	167.5	2,843	315.9	39.6
	Waste[2] (g)	418	171	—	5,850	1,800
Enough bags to carry 483 items[3]	Global warming (kg CO_{2eq})	1.578	6.924	5.523	21.51	271.5
	Acidification (g SO_{2eq})	11.4	29.3	37.5	101.3	2,788
	Eutrophication (g $PO_4^{3-}{}_{eq}$)	0.775	2.58	5.04	14.6	304
	Human toxicity (kg $1,4\text{-}DB_{eq}$)	0.211	0.701	3.25	3.05	66.3
	Freshwater ecotoxicity (g $1,4\text{-}DB_{eq}$)	66.9	187	150	468	23,480

[1]Virgin fibers are necessary to provide strength.
[2]Waste generated during manufacture.
[3]In the United Kingdom, 483 items make the average monthly grocery shopping.
Source: [105] Tables 1.1, 4.2, 5.1, 5.4, 5.5, 5.6, and 5.7 with some ranges of numbers reduced to their average.

The conclusion is that the paper bag needs to be used at least 4 times, the reusable PP bag 14 times, and the cotton bag 173 times ([105] page 33) to have a carbon footprint lower than that of the light-plastic (HDPE) bag.

12.11 Service industry

Since it does not manufacture anything but only uses office and commercial buildings as well as some transportation, the service industry has a relatively light environmental footprint relative to its importance in the overall economy: In a developed country, its greenhouse gas emissions are less than 12% [59] while its GDP contribution is over 50% [106].

One way to assess the environmental footprint of the service sector is to determine the impacts caused by an employee at work. Seventeen independent life-cycle assessments were reviewed [107], with the conclusions summarized in the following table. The lowest and highest carbon footprints among the seventeen companies were, respectively, 3.1 and 21 metric tons of CO_{2eq} per employee per year, with median and average impacts of 5.9 and 7.4 metric tons of CO_{2eq} per employee per year ([107] page 24). The author's own analysis for office buildings in Finland, where energy is somewhat greener, came a little lower at 4.92 metric tons of CO_{2eq} per employee per year ([107] Section 7.4.2). For comparison, Google states that its carbon footprint per full-time employee is 8.4 metric tons of CO_{2eq} per year ([108] Fig. 11).

Carbon footprint per employee in the service sector				
Element or Activity	Low	Median	High	Units
Office space	1,426	2,922	7,000	
Commuting and business travel	54	740	6,740	
Office supplies	30	100	638	kg CO_{2eq} per employee per year
Office equipment	62	400	1,300	
Purchased services	120	1,000	10,200	

Source: [107] Section 3.2 (also Fig. 34).

12.11.1 Travel, tourism, and hospitality

Worldwide, nearly 1.3 billion people traveled in 2017 for business or pleasure [109]. Aside from times of economic recession, the tourism industry represents 5% of world GDP while contributing to about 8% of total employment ([110] page vii). It also contributes 5% of global carbon dioxide emissions, with transport being responsible for over 90% chiefly because of air travel ([110] pages 3–4). For a short holiday, the majority (60–95%) of the carbon footprint is caused by air travel, if any [111].

Evidently, some people engage in leisure travel more than others as the following table reveals.

Share of international tourist arrivals by region				
Europe	Asia and Pacific	Americas	Africa	Middle East
52%	21%	16%	6%	6%

Source: [110] Fig. 3.4.

The following table provides estimated impacts of global tourism.

Various impacts of global tourism, per tourist			
Aspect	Range	Global average	Units
Energy			
Domestic and international travel	50–136,000	3,575	MJ/trip
Accommodation	3.6–3,717	272	MJ/night
Carbon emissions			
Domestic and international travel	<100–9,300	250	kg CO_2/trip
Accommodation	0.1–260	13.8	kg CO_2/ night
Freshwater			
Accommodation	84–2,425	350	L/day
Indirect for food and fuels	4,500–8,000	6,000	L/day
Combined	4,600 –12,000	6,575	L/day
Food			
Meals	1.28–3.1	1.8	kg/day
Land use			
Direct	30–4,580	42	m^2/bed
Infrastructure	–	11.7	m^2

Source: [112] Table 5.

Tourism inevitably generates waste, much of it in the form of litter in otherwise scenic environments. The United Nations Environment Programme estimates that each year tourists generate 4.8 million tons of waste, 14% of which is solid waste. Of the 12.2 million tons of plastic entering the ocean every year, it is estimated that over 80% come from land-based sources, with seashore litter from drink bottles and other packaging being the largest contributor [113].

Cruise ships are notorious polluters. The US Bureau of Transportation reports that a typical one-week voyage with 3,000 passengers and crew on board the ship at sea generates 210,000 gallons (795 m^3, a third of an Olympic swimming pool) of sewage, 8 short tons (7.2 metric tons) of solid waste (mostly plastics and food waste), 130 gallons (490 L) of hazardous waste (mostly photochemicals and used paint), and 25,000 gallons (95,000 L) of so-called oily bilge water (water collected at the bottom of the ship) [114]. In the Caribbean Sea, waste brought onshore by cruise ships during port calls amount to 3.5 kg of garbage per passenger per day, a sizeable amount compared with the 0.8 kg per person per day generated by the less affluent local population [115].

International tourism across Europe generates at least 1 kg of solid waste per tourist per day, with tourists from the United States found to produce even more, up to 2 kg daily per tourist. It was estimated that the world's 692.5 million international tourists during 2001 had generated at least 4.8 million tons of solid waste (58% of it in Europe alone), giving an average of 6.9 kg of solid waste per tourist. When domestic tourism is added, the global figure for solid waste rises to 35 million tons per year, which is about the same as the solid waste generated in all of France every year ([116] page 3).

In the United States, tourism and recreation consume 946 million m^3 of water per year, of which 60% is related to lodging (mostly spent on hotel guest consumption, property management including landscaping, and laundry) and another 13% to foodservice. In Europe, it is estimated that every tourist consumes 300 L (80 gallons) of fresh water per day. Luxury resorts typically use larger volumes of water, up to 800 L (210 gallons) daily per guest ([116] page 21).

In a tropical country such as Thailand, the average golf course uses 1,500 kg of fertilizers, pesticides, and herbicides, and its watering consumes as much water as 60,000 rural villagers [115].

The footprint of a tourist staying for an average of 3.2 days on Penghu Island in Taiwan is estimated at 1,606 MJ of energy, 607 L of water, 416 L of wastewater with 83.1 g of BOD, 109 kg of CO_2, 2.66 kg of CO, 597 g of hydrocarbons, 70 g of nitrogen oxides, and 1.95 g of solid waste (per tourist per stay). In terms of relative energy use, transportation constitutes the largest share (67%), due to air travel. Needless to say, a tourist to Penghu Island causes an environmental load exceeding that of local people; as one example, the 1.95 kg/day of solid waste generated by a tourist is greater than the 1.18 kg/day of a local person [117].

The estimated global average energy use by type of accommodation is as follows:

Footprint for different types of overnight accommodations		
Type of accommodation	Energy use per guest-night (MJ)	Emission per guest-night (kg CO_2)
Hotel	130	20.6
Campground	50	7.9

Continued

Pension	25	4.0
Self-catering	120	19.0
Holiday village	90	14.3
Vacation home	100	15.9
AVERAGE	98	15.6

Source: [110] Table 2.3.

12.11.2 Retail

A comparison between in-store and on-line retail can be found in Section 8.3.5. Section 2.3 provides additional numbers on water and wastewater.

The following are characteristics of brick-and-mortar retail stores, compiled from a set of more than 55,000 properties across the United States.

Characteristics of retail stores			
Characteristic	**Range**	**Median**	**Units**
Floor size	7,106 −127,455	14,010	ft^2
Hours	71−168	91	hours/week
Workers	0.1−1.0	0.4	workers/1,000 ft^2
Cash registers	0.1−0.8	0.3	units/1,000 ft^2
Computers	0.0−0.8	0.2	units/1,000 ft^2
Energy use	69−400	192	kBTUs/ft^2
Open/closed refrigeration cases	0.0−1.0	0.2	units/1,000 ft^2

Source: [118].

Water intensity[3] is about 12.5 gallons/ft^2/yr for detached stores, 11.7 gallons/ft^2/yr in shopping malls, and 3.4 gallons/ft^2/yr in warehouses, all of which fall below the average of 20.3 gallons/ft^2/yr for commercial buildings ([73] Fig. 1). The relative water consumption by end-use is tabulated in the following, with a comparison to restaurants.

---

3. For conversion to metric units: 1 gallon/ft^2/yr = 40.75 L/m^2/yr.

Water consumption in retail stores vs. restaurants		
End use	**Average retail store**	**Restaurants**
Cooling	27%	1.5%
Restrooms	24%	34%
Landscaping	31%	6.5%
Kitchen	5%	46%
Other	13%	12%
Source: [2] Fig. 4-1.		

The following table gives the daily water consumption for a selection of stores.

Water consumption in selected stores			
Store type	**gallons/day**	**Store type**	**gallons/day**
Auto-repair shops	592	Auto service stations	1,683
Food services	4,480	Grocery stores	7,490
Florists	6,277	Shopping centers/malls	7,083
Funeral services	640	Retail stores	1,353
Source: [119] Table 2.			

Water consumption in grocery stores varies between 2.6 and 4.5 gallons per transaction [120].

12.12 Textile industry

The following table lists the greenhouse gas emissions in the production of the most common fibers used in the textile industry.

Carbon footprint of fibers		
Textile material	**kg CO_{2eq}/kg**	**Source**
Average of common textiles	3.89	[23] p.112
Acrylic	26	[121] Table 7.7
	7.6	[122] Table 1

Continued

Carbon fiber	23.9–26.4	[123] p. 573
Coir (from coconut shells)	0.427 –0.472	[123] p. 577
Cotton[1]	3.47	incl. ginning [124] Table 1.2
	2.4–2.7	[123] p. 579
	1.76	[122] Table 1
Felt	0.96	[23] p.112
Fiberglass	3.34–3.69	[123] p. 575
Flax/linen	0.37–0.41	[123] p. 581
	0.34	[122] Table 1
Hemp	0.29–0.33	[123] p. 583
Jute	1.2–1.4	[123] p. 585
Kenaf	1.3–1.4	[123] p. 587
Kevlar	82.1–90.8	[123] p. 571
Neoprene	1.6–1.8	[123] p. 525
Nylon	8.00	[124] Table 1.2
	5.43	[23] p. 112
Polyamide	8.1	[122] Table 1
Polyester	2.8–3.2	[123] p. 515
	5.4	[122] Table 1
Polyurethane/ Polypropylene	3.1	[122] Table 1
Ramie	0.43–0.47	[123] p. 589
Silk	52.50	Incl. all farming operations [124] Table 1.2
	2.03	[122] Table 1
Sisal (from agave plant)	0.42–0.47	[123] p. 591
Viscose	2.1	[122] Table 1
Wool	3.2–3.5	[123] p. 593
	5.48	[23] p.112
	20.8	[122] Table 1

Caveat: Numbers are hardly comparable as different sources made different assumptions about the boundaries of their life-cycle assessments. The reader is referred to the original sources for details.
[1]*The major environmental impact of cotton production is water usage; see numbers below. The discrepancy in carbon footprints listed here is indicative of different assumptions made about the boundaries of the analysis such as whether the energy for irrigation is included or not and how much transportation is involved.*

The daily water consumption of a textile mill producing about 20,000 lbs (9,100 kg) of fabric per day is 36,000 L (9,500 gal) [125]. It takes about 20 gallons (76 L) of water to produce 1 yard (0.91 m) of upholstery fabric; so, if there are about 25 yards (23 m) of upholstery fabric on a sofa, the water necessary to produce the fabric to cover that one sofa is 500 gallons (1,900 L) [126].

Water use by the textile industry has been documented by the processing category, as tabulated in the following.

Type of fabric	Range (gal/lb)	Median (gal/lb)
Wool	13.3−78.9	34.1
Felted fabric	4.0−111.8	25.5
Woven	0.6−60.9	13.6
Stock/yarn	0.4−66.9	12.0
Knit	2.4−45.2	10.0
Carpet	1.0−19.5	5.6
Non-woven	0.3−9.9	4.8

Source: [127] Table F-20.

Water is consumed in the course of several consecutive processes: 15% preparation, 52% dyeing (incl. multiple rinses), 6% printing, and 27% final washing ([127] Table F-22).

Dyeing of yarn and fabric is water and chemical-intensive. Water consumption for dyeing is about 60 L per kg of yarn and between 30 and 50 L per kg of fabric, depending on the type of dye used. With about 80% of the dye staying on the fabric or yarn, the 20% remainder finds its way into wastewater [128]. As a result, wastewater from textile mills is heavily polluted as numbers show in the following table. A total of 72 toxic chemicals have been identified in effluents from textile dyeing, 30 of which cannot be removed [128].

Typical composition of effluent water from a textile mill			
Characteristic	Concentration	Characteristic	Concentration
pH	9.2−11	COD	465−1,400 mg/L
Alkanility	1,250−3,160 mg/L	Total dissolved solids	3,230−6,180 mg/L
BOD	130−820 mg/L	Total suspended solids	360−370 mg/L

Source: [128].

The amounts of energy used and carbon footprint in fiber production are given in the following.

Energy use and carbon footprint in fiber production		
Fiber	Energy use (MJ/kg of fiber)	Carbon footprint (CO_{2eq} relative to wool)[1]
Acrylic	155–160	0.36
Nylon	135–140	0.32
Polyester	105–110	0.23
Cotton	45–50	0.18
Wool	40–45	1.00
Viscose	30–35	0.13
Linen	20–25	0.09

[1]These numbers are relative to one type of fiber because the carbon footprint depends on the cleanliness of energy, which in turn varies widely on the region of production, from 1 kg CO_{2eq}/kWh in a coal-fired power plant in China down to zero for a hydroelectric plant in Brazil. Wool has the highest global warming potential because of methane emissions from the sheep (contribution of 85% to the total).
Source: [129] Charts 2a and 2b.

Cotton is the most important natural fiber in the apparel sector, accounting for about 40% of all textile production, after synthetic fibers that account for 55%. Its production is the most water-intensive segment in the apparel sector and, as so, is also a sector vulnerable to climate-induced water risks. About 53% of all cotton fields are irrigated, producing more than their proportional share (73%) of production [130]. The water need is 3,644 L/kg, of which 1,827 L is supplied by rainwater ("green water") and 1,818 L by irrigation ("blue water" from surface and subsurface sources) ([130] Table 3.4). And, this is not all; additional water is required to process the cotton into fabric. For the final cloth, 4,917 L/kg of nonprecipitation water is required ([130] Table 3.5). This implies that 1,230 L are needed to produce the 250 g of cotton for the humble 170-g T-shirt.[4]

The environmental impacts of the average 170-g cotton T-shirt are tabulated in the following.

---

4. The sometimes quoted higher value of 25 m^3 for the cotton T-shirt includes the natural precipitation water component, which about doubles the footprint.

Environmental impacts of a T-shirt	
Type of impact	Impact
Abiotic depletion	1.08 g Sb_{eq}
Acidification	0.777 g SO_{2eq}
Eutrophication	0.3771 g PO_{4eq}
Global warming potential (GWP100)	139.41 g CO_{2eq}
Ozone layer depletion	1.26×10^{-8} kg CFC-11 eq
Human toxicity[1]	63.65 g 1,4 DB_{eq}
Freshwater aquatic ecotoxicity	62.17 g 1,4 DB_{eq}
Marine aquatic ecotoxicity	134.9 kg 1,4 DB_{eq}
Terrestrial ecotoxicity	0.955 g 1,4 DB_{eq}

[1]Toxicity is measured in g or kg of equivalent 1,4 dichlorobenzene (noted 1,4 DB_{eq}).
Source: [131] Table 1.

The next table lists the energy consumption in the various stages of making a 170-g cotton T-shirt, following cotton cloth manufacture and excluding transportation.

Energy consumption in the manufacture of a T-shirt		
Process	Energy consumption	Fraction
Cutting	0.732 MJ	29.6%
Sewing	1.23 MJ	49.8%
Packaging	0.51 MJ	20.6%
Total	2.472 MJ	100%

Source: [131] Table 2.

Besides 250 g of cotton, it takes 2,000 L (528 gallons) of water, 150 g (1/3 lb) of chemicals, 23,500 km (14,600 mi) of air travel from the United States to China and back to make a 170-g T-shirt. Similarly, a pair of jeans typically travels 20,000 miles before sold in the United States [48].

The following are the breakdowns of carbon footprints for the production of a cotton T-shirt and of a multimaterial jacket.

Carbon footprint of a T-shirt	
Component or process	**kg CO_{2eq}**
Cotton cultivation and ginning	0.26
Yarn spinning	0.45
Weaving	0.41
Bleaching	0.19
Drying	0.19
Sewing	0.34
Total	1.84

Source: [124] Fig. 1.4.

Carbon footprint of a jacket	
Component or process	**kg CO_{2eq}**
Polyamide fibers	2.3
Weave of polyamide fibers	3.9
Polyester fibers	0.5
Weave of polyester fibers	0.6
Polyester padding	0.8
Nonwoven processes	0.6
Cotton fibers	0.7
Cotton tricot production	1.1
Garment manufacturing	1.9
Total	12.4

Source: [124] Fig. 1.3.

Sources

[1] P.W. Griffin, G.P. Hammond, J.B. Norman, Industrial energy use and carbon emissions reduction in the chemicals sector: a UK perspective, Appl. Energy 227 (2018) 587–602. doi.org/10.1016/j.apenergy.2017.08.010.

[2] P.H. Gleick, et al., Pacific Institute — Waste Not, Want Not: The Potential For Urban Water Conversation in California, November 2003, 165 pages. pacinst.org/wp-content/uploads/2003/11/waste_not_want_not_full_report.pdf.

[3] University of Oxford — Oxford Martin School, Our World in Data — Land Use, by Hannah Ritchie and Max Roser, September 2019. ourworldindata.org/land-use.

[4] A.Y. Hoekstra, M.M. Mekonnen, The water footprint of humanity, Proc. Nat. Acad. Sci. U.S.A 109 (9) (2012) 3232–3237. doi.org/10.1073/pnas.1109936109.

[5] United Nations − Food and Agriculture Organization (FAO), FAO Newsroom − Livestock a Major Threat to Environment, November 29, 2006. www.fao.org/newsroom/en/news/2006/1000448/index.html.

[6] Yara International ASA (Norway), Crop nutrition solutions − Fertilizer life cycle perspective, undated posting. www.yara.com/crop-nutrition/why-fertilizer/environment/fertilizer-life-cycle/.

[7] World Bank, Data − Fertilizer Consumption (Kilograms Per Hectare of Arable Land), (2016 data). data.worldbank.org/indicator/AG.CON.FERT.ZS?end=2016&start=2002&view=chart.

[8] N. Anderson, R. Strader, C. Davidson, Airborne reduced nitrogen: ammonia emissions from agriculture and other sources, Environ. Int. 29 (2−3) (2003) 277−286. doi.org/10.1016/S0160-4120(02)00186-1.

[9] M.M. Mekonnen, A.Y. Hoekstra, The green, blue and grey water footprint of farm animals and animal products, in: Main Report. Value of Water Research, Report 48, vol. 1, UNESCO − IHE Institute for Water Education, Delft, the Netherlands, 2010, 50 pages.

[10] University of Oxford − Oxford Martin School, Our World in Data − Meat and Dairy Production, Hannah Ritchie and Max Roser, November 2019. ourworldindata.org/meat-production.

[11] University of Oxford − Oxford Martin School, Our World in Data − Crop Yields, in: Hannah Ritchie and Max Roser, September 2019. ourworldindata.org/crop-yields.

[12] C.L. Weber, H.S. Matthews, Food-miles and the relative climate impacts of food choices in the United States, Environ. Sci. Technol. 42 (2008) 3508−3513. doi.org/10.1021/es702969f.

[13] R.C. Taylor, H. Omed, G. Edwards-Jones, The greenhouse emissions footprint of free-range eggs, Poult. Sci. 93 (1) (2014) 231−237. doi.org/10.3382/ps.2013-03489.

[14] M. MacLeod, P. Gerber, A. Mottet, G. Tempio, A. Falcucci, C. Opio, T. Vellinga, B. Henderson, H. Steinfeld, Greenhouse Gas Emissions from Pig and Chicken Supply Chains − A Global Life Cycle Assessment, Food and Agriculture Organization of the United Nations (FAO), Rome, 2013, 171 pages.

[15] C. Opio, P. Gerber, A. Mottet, A. Falcucci, G. Tempio, M. MacLeod, T. Vellinga, B. Henderson, H. Steinfeld, Greenhouse Gas Emissions from Ruminant Supply Chains − A Global Life Cycle Assessment, Food and Agriculture Organization (FAO) of the United Nations, Rome, 2013, 191 pages.

[16] P. Gerber, T. Vellinga, C. Opio, B. Henderson, H. Steinfeld, Greenhouse Gas Emissions from the Dairy Sector − A Life Cycle Assessment, Food and Agriculture Organization (FAO) of the United Nations, Rome, 2010, 94 pages.

[17] C. Purdy, Need an excuse to avoid broccoli? Point to carbon emissions. Quartz, web posting dated December 31, 2019. qz.com/1777399/by-calorie-broccoli-is-a-bigger-carbon-emitter-than-chicken/.

[18] M.M. Mekonnen, A.Y. Hoekstra, The green, blue and grey water footprint of crops and derived crop products, Hydrol. Earth Syst. Sci. 15 (2011) 1577−1600. doi.org/10.5194/hess-15-1577-2011.

[19] F. Scrucca, E. Bonamente, S. Rinaldi, Carbon footprint in the wine industry. Chap. 7 in: S.S. Muthu (Ed.), Environmental Carbon Footprints − Industrial Case Studies, Butterworth-Heinemann, Elsevier, 2018, pp. 161−196. doi.org/10.1016/B978-0-12-812849-7.00007-6.

[20] C. Foster, K. Green, M. Bleda, P. Dewick, B. Evans, A. Flynn, J. Mylan, Environmental Impacts of Food Production and Consumption: A Report to the Department for Environment, Food and Rural Affairs (Defra, London), Manchester Business School, 2006, 199 pages, leanenterprise.org.uk/wp-content/uploads/2018/12/DEFRA-SCP007-ENVIRONMENTAL-IMPACTS-OF-FOOD-CONSUMPTION-AND-PRODUCTION.pdf.

[21] E.M. Bell, A. Horvath, Modeling the carbon footprint of fresh produce: effects of transportation, localness, and seasonality on US orange markets, Environ. Res. Lett. 15 (2020) 034040. doi.org/10.1088/1748-9326/ab6c2f.

[22] R.C. Fluck, Energy for Florida Oranges. Fact Sheet EES-81, Florida Cooperative Extension Service, University of Florida, 1992. October 1992, 3 pages, ufdcimages.uflib.ufl.edu/IR/00/00/48/24/00001/EH18100.PDF.

[23] M. Berners-Lee, How Bad Are Bananas? The Carbon Footprint of Everything, Greystone Books, D&M Publishers, 2011, 232 pages.

[24] I. Randall, Butter produces three-and-half times more greenhouse gases than margarine — and burping and farting cows are to blame, DailyMail, 2020. March 10, 2020, www.dailymail.co.uk/sciencetech/article-8095901/Gassy-cows-blame-butter-revealed-produce-3-5-times-carbon-dioxide-margarine.html.

[25] S. Becker, T. Bouzdine-Chameeva, A. Jaegler, The carbon neutrality principle: a case study in the French spirits, J. Clean. Prod. 274 (2020) 122739. doi.org/10.1016/j.clepro.2020.122739.

[26] S. Poulikidou, Identification of the Main Environmental Challenges in a Sustainability Perspective for the Automobile Industry, M.S. Thesis, Dept. Energy and Environment, Chalmers University of Technology, 2010, 71 pages, publications.lib.chalmers.se/records/fulltext/136380.pdf.

[27] U.S. Department of Energy, Argonne National Laboratory — Energy Systems Division — End Of-Life Vehicle Recycling: State of the Art of Resource Recovery from Shredder Residue, Report ANL/ESD/10-8, September 2010, 163 pages, publications.anl.gov/anlpubs/2011/02/69114.pdf.

[28] Organisation Internationale des Constructeurs d'Automobiles (OICA) —, Production Statistics, 2019. www.oica.net/category/production-statistics/2019-statistics/.

[29] Organisation Internationale des Constructeurs d'Automobiles (OICA), Motorization Rate, 2015. www.oica.net/category/vehicles-in-use/.

[30] European Automobile Manufacturers Association, The Automobile Industry Pocket Guide 2020-2021, 88 pages. www.acea.be/publications/article/acea-pocket-guide.

[31] The International Council on Clean Transportation (ICCT), Briefing — Effects of Battery Manufacturing on Electric Vehicle Life-Cycle Greenhouse Gas Emissions, February 2018, 12 pages, theicct.org/sites/default/files/publications/EV-life-cycle-GHG_ICCT-Briefing_09022018_vF.pdf.

[32] T.R. Hawkins, B. Singh, G. Majeau-Bettez, A.H. Strømman, Comparative environmental life cycle assessment of conventional and electric vehicles, J. Ind. Ecol. 17 (1) (2013) 53—64. doi.org/10.1111/j.1530-9290.2012.00532.x.

[33] University of Michigan, Center for Sustainable Systems — Personal Transportation Factsheet, Pub. No. CSS01-07, August 2019. css.umich.edu/factsheets/personal-transportation-factsheet.

[34] I.Z. Bribián, A.V. Capilla, A.A. Usón, Life cycle assessment of building materials: comparative analysis of energy and environmental impacts and evaluation of the eco-efficiency improvement potential, Build. Environ. 46 (2011) 1133—1140. doi.org/10.1016/j.buildenv.2010.12.002.

[35] R.H. Crawford, G.J. Treloar, An assessment of the energy and water embodied in commercial building construction, in: Proc. 4th Australian Life-Cycle Assessment Conf., February 2005, Sydney, 2005, 10 pages, www.academia.edu/650424/Crawford_R.H._and_ Treloar_G.J._2005_An_assessment_of_the_energy_and_water_embodied_in_ commercial_building_construction_Proceedings_4th_Australian_Life-Cycle_Assessment_ Conference_Sydney_February_10p.

[36] U.S. Energy Information Administration, Use of Energy Explained − Energy Use in Industry − Basics. www.eia.gov/energyexplained/use-of-energy/industry.php.

[37] World Economic Forum, Davos 2020 - How to Build a More Climate-Friendly Chemical Industry − Background Paper by Martin Brudermüller, January 21, 2020. www.weforum. org/agenda/2020/01/how-to-build-a-more-climate-friendly-chemical-industry/.

[38] Organization for Economic Co-operation and Development (OECD), Environmental Outlook for the Chemicals Industry, 2001, 164 pages, www.oecd.org/env/ehs/2375538.pdf.

[39] A.G. Bayer, Annual Report, 2019, 239 pages, bayer-ag-annual-report-2019.pdf.

[40] Energy and carbon numbers from pages 47-51 in: One Linde − Sustainable Development Report 2018, 101 pages. www.linde.com/-/media/linde/merger/documents/sustainable-development/2018-sustainable-development-report.pdf? rev=4b91ad8384b74e10b2304aca96022c4a.

[41] U. S. Environmental Protection Agency, Toxics Release Inventory (TRI) - 2018 National Analysis − Executive Summary, 5 pages. www.epa.gov/sites/production/files/2020-02/ documents/tri_national_analysis_executive_summary.pdf.

[42] U. S. Environmental Protection Agency, Toxics Release Inventory (TRI) - 2018 National Analysis − Comparing Industry Sectors, February 2020, 42 pages, www.epa.gov/sites/ production/files/2020-02/documents/industry_sectors.pdf.

[43] Wikipedia, List of industrial Disasters. en.wikipedia.org/wiki/List_of_industrial_disasters.

[44] Panels & Furniture, Group of Wood Magazines - UN, Wooden furniture reduces more carbon emissions than other materials, posted on 25 July 2016, quoting a report by the U.N. Food and Agriculture Organization. www.panelsfurnitureasia.com/en/news-archive/un-wooden-furniture-reduces-more-carbon-emissions-than-other-materials/462.

[45] A. Druckman, T. Jackson, An Exploration into the Carbon Footprint of UK Households, RESOLVE Working Paper 02-10, Centre for Environmental Strategy (D3), University of Surrey, 2010, 35 pages, resolve.sustainablelifestyles.ac.uk/sites/default/files/RESOLVE_ WP_02-10.pdf.

[46] Furniture Industry Research Association (FIRA), Benchmarking Carbon Footprints of Furniture Products, April 2011, 60 pages, www.fira.co.uk/images/fira-carbon-footprinting-document-2011.pdf.

[47] S. Gajalakshmi, T. Abbasi, P.S. Ganesh, S.A. Abbasi, Rapid treatment of shampoo industry waste to significantly reduce capital and operational costs, J. IPHE India 2006−07 (4) (2006) 25−27. www.bits-pilani.ac.in/uploads/PSGHandouts/Shampoo.pdf.

[48] Catskill Mountainkeeper, Human Footprint − Journey of a Life Time. www. catskillmountainkeeper.org/human_foot_print.

[49] H. Haroon, A. Waseem, Q. Mahmood, Treatment and reuse of wastewater from beverage industry, J. Chem. Soc. Pak. 35 (1) (2013) 5−10.

[50] The Coca Cola Company, Improving Our Water Efficiency − Meeting Goals and Moving Goalposts, posted 29 August 2018. www.coca-colacompany.com/news/improving-our-water-efficiency.

[51] A.K. Chapagain, A.Y. Hoekstra, Water Footprints of Nations. UNESCO Value of Water Research Report Series No. 16, 2004. November 2002, 80 pages, waterfootprint.org/media/downloads/Report16Vol1.pdf.

[52] Statista — Agriculture — Area of harvested tobacco worldwide from 1980 to 2018. www.statista.com/statistics/261192/global-area-of-harvested-tobacco-since-1980/.

[53] H.J. Geist, Global assessment of deforestation related to tobacco farming, Tob. Control 8 (1) (1999) 18—28. doi.org/10.1136/tc.8.1.18. Available from: tobaccocontrol.bmj.com/content/8/1/18.full.

[54] The Tobacco Atlas, Consumption. tobaccoatlas.org/topic/consumption/.

[55] N.S. Hopkinson, D. Arnott, N. Voulvoulis, Environmental consequences of tobacco production and consumption, Lancet 394 (2019) 1007—1008. doi.org/10.1016/S0140-6736(19)31888-4. Data in supplement available at: www.thelancet.com/cms/10.1016/S0140-6736(19)31888-4/attachment/0063df22-0b46-42fc-9513-cb4954c42cb7/mmc1.pdf.

[56] World Health Organization, Policy and Practice - the environmental and health impacts of tobacco agriculture, cigarette manufacture and consumption, Bulletin 93 (2015) 877—880. doi.org/10.2471/BLT.15.152744, posted on 22 October 2015, www.who.int/bulletin/volumes/93/12/15-152744/en/.

[57] Lawrence Livermore National Laboratory, Energy Flow Chart, 2018. flowcharts.llnl.gov/content/assets/docs/2018_United-States_Energy.pdf.

[58] U.S. Energy Information Administration, Electric Generation Transforms Primary Energy into Secondary Energy — U.S. Electricity Flow, 2018. www.eia.gov/todayinenergy/detail.php?id=41193.

[59] U.S. Environmental Protection Agency, Greenhouse Gas Emissions — Sources of Greenhouse Gas Emissions, (2018 data). www.epa.gov/ghgemissions/sources-greenhouse-gas-emissions.

[60] U.S. Energy Information Administration, Today in Energy — Carbon Intensity of Energy Use Is Lowest in U.S. Industrial and Electric Power Sectors, (2016 data). www.eia.gov/todayinenergy/detail.php?id=31012.

[61] Intergovernmental Panel on Climate Change (IPCC), Fifth Assessment Report, 2014. Chapter 7: Energy Systems, authored by Thomas Bruckner et al., 88 pages, www.ipcc.ch/site/assets/uploads/2018/02/ipcc_wg3_ar5_chapter7.pdf.

[62] E.S. Spang, W.R. Moomaw, K.S. Gallagher, P.H. Kirshen, D.H. Marks, The water consumption of energy production: an international comparison, Environ. Res. Lett. 9 (10) (2014) 105002. doi.org/10.1088/1748-9326/9/10/105002.

[63] A.J. Kondash, D. Patino-Echeverri, A. Vengosh, Quantification of the water-use reduction associated with the transition from coal to natural gas in the US electricity sector, Environ. Res. Lett. 14 (2019) 124028. doi.org/10.1088/1748-9326/ab4d71.

[64] U. Lee, J. Han, A. Elgowainy, M. Wang, Regional water consumption for hydro and thermal electricity generation in the United States, Appl. Energy 210 (2017). doi.org/10.1016/j.apenergy.2017.05.025.

[65] U.S. Energy Information Administration, Frequently Asked Questions (FAQs) - How Much Carbon Dioxide Is Produced Per Kilowatthour of U.S. Electricity Generation? www.eia.gov/tools/faqs/faq.php?id=74&t=11.

[66] World Nuclear Association, Comparison of Lifecycle Greenhouse Gas Emissions of Various Electricity Generation Sources, 2011, 12 pages. www.world-nuclear.org/uploadedFiles/org/WNA/Publications/Working_Group_Reports/comparison_of_lifecycle.pdf.

[67] U.S. Department of Energy, Comparative Life-Cycle Air Emissions of Coal, Domestic Natural Gas, LNG, and SNG for Electricity Generation, 2013, 831 pages. fossil.energy.gov/ng_regulation/sites/default/files/programs/gasregulation/authorizations/2013/applications/SC_Ex_13_04_lngPart_4.pdf.

[68] International Hydropower Association, Climate Change − Greenhouse Gas Emissions − Hydropower Is a Low-Carbon Technology Which Helps to Mitigate the Carbon Emissions of Fossil Fuels. www.hydropower.org/greenhouse-gas-emissions.

[69] E.G. Hertwich, Addressing biogenic greenhouse gas emissions from hydropower in LCA, Environ. Sci. Technol. 47 (2013) 9604−9611. doi.org/10.1021/es401820p.

[70] T. Davies, A.I. Lowe, Environmental Implications of the Health Care Service Sector. Discussion Paper 00-01, Resource for the Future, Washington DC, October 1999, 1999, 40 pages. www.rff.org/publications/working-papers/environmental-implications-of-the-health-care-service-sector/.

[71] U.S. Energy Information Administration, Consumption & Efficiency − 2012 Commercial Buildings Energy Consumption Survey (CBECS) − Energy Usage Summary, March 2016. www.eia.gov/consumption/commercial/reports/2012/energyusage/index.php.

[72] U.S. Energy Information Administration, Consumption & Efficiency − 2012 Commercial Buildings Energy Consumption Survey (CBECS) − Table PBA4. Electricity Consumption Totals and Conditional Intensities by Building Activity Subcategories, December 2016. www.eia.gov/consumption/commercial/data/2012/c&e/cfm/pba4.php.

[73] U.S. Energy Information Administration, Commercial Buildings Energy Consumption Survey (CBECS 2012) − Water Consumption in Large Buildings Summary, February 9, 2017. www.eia.gov/consumption/commercial/reports/2012/water/.

[74] C.C. Lee, G.L. Huffman, Medical waste management/incineration, J. Hazard. Mater. 48 (1−3) (1996) 1−30. doi.org/10.1016/0304-3894(95)00153-0.

[75] U.S. Environmental Protection Agency, Office of Air Quality Planning and Standards − Emission Factor Documentation for AP-42 Section 2.6, reformatted 1/95, 217 pages. www3.epa.gov/ttnchie1/ap42/ch02/bgdocs/b02s03.pdf.

[76] U.K. National Health Service, Sustainable Development Unit − Policy and Strategy − Natural Resource Footprint − 2018 Report, 31 pages. www.sduhealth.org.uk/policy-strategy/reporting/natural-resource-footprint-2018.aspx.

[77] J.H. Clark, Chemistry of Waste Minimization, Springer, 1995, 554 pages.

[78] P.J. Dunn, R.K. Henderson, I. Mergelsberg, A.S. Wells, Moving towards Greener Solvents for Pharmaceutical Manufacturing − an Industry Perspective. 13th Annual Green Chemistry & Engineering Conference, 2009, pp. 23−25. June 2009, College Park, Maryland.

[79] A. Fairbanks, Energy Efficiency in Biotech and Pharma. Fairbanks Energy Services, posted 3 August 2020, 2020. blog.fairbanksenergy.com/energy-efficiency-in-biotech-and-pharma.

[80] Arcadis − Optimizing Operational Expenditure − A More Effective Approach than Just Cost Reduction, Circa, 2012, 8 pages. www.arcadis.com/media/6/2/8/%7B6282C3E1-D470-4BF0-B312-C5EE8309A3AF%7DOptimizing%20Operational%20Expenditure.pdf.

[81] L. Belkhir, A. Elmeligi, Carbon footprint of the global pharmaceutical industry and relative impact of its major players, J. Clean. Prod. 214 (2019) 185−194. doi.org/10.1016/j.jclepro.2018.11.204.

[82] ezeep, ThinPrint GmbH, What Is Printing's Carbon Footprint? − © 2020. www.ezeep.com/co2-neutral-printing/.

[83] H. Pihkola, M. Nors, M. Kujanpää, T. Helin, M. Kariniemi, T. Pajula, Carbon Footprint and Environmental Impacts of Print Products from Cradle to Grave, VTT Tiedotteita − Research Notes 2560, 2010. Helsinki, 253 pages. www.vttresearch.com/sites/default/files/pdf/tiedotteet/2010/T2560.pdf.

[84] Heidelberger Druckmaschinen AG, Offsetting the Carbon Footprint of Your Printing Press, 2020. www.heidelberg.com/global/en/products/co2_neutral_certificate/CO2_neutral_equipment_certificate.jsp.

[85] C.J. Corbett, R.P. Turco, Sustainability in the Motion Picture Industry, Institute of the Environment, University of California Los Angeles (UCLA), 2006, 114 pages. www.ioes.ucla.edu/wp-content/uploads/mpisreport.pdf.

[86] British Academy of Film and Television Arts (BAFTA) — Albert Annual Report, What Goes up Must Come Down, 2018, 12 pages. static.bafta.org/files/albert-year-one-report-carbon-footprinting-the-tv-industry-1574.pdf.

[87] C.L. Weber, J.G. Koomey, H.S. Matthews, The energy and climate change impacts of different music delivery methods, J. Ind. Ecol. 14 (5) (2010) 754–769. doi.org/10.1111/j.1530-9290.2010.00269.x.

[88] C. Bottrill, D. Liverman, M. Boykoff, Carbon soundings: greenhouse gas emissions of the UK music industry, Environ. Res. Lett. 5 (2010) 014019. doi.org/10.1088/1748-9326/5/1/014019, 8 pages, sciencepolicy.colorado.edu/admin/publication_files/2010.13.pdf.

[89] G. Karras, Oil Refinery CO_2 Performance Measurement. Technical Analysis Prepared for the Union of Concerned Scientists, File No. COMMBETTERENVFY 11103, 2011. September 2011, 20 pages. insideclimatenews.org/sites/default/files/ucs-report-oil-refinery-CO2-performanceINSIDECLIMATENEWS.pdf.

[90] M.S. Masnadi, H.M. El-Houjeiri, D. Schunack, Y. Li, J.G. Englander, A. Badahdah, J.-C. Monfort, J.E. Anderson, T.J. Wallington, J.A. Bergerson, D. Gordon, J. Koomey, S. Przesmitzki, I.L. Azevedo, X.T. Bi, J.E. Duffy, G.A. Heath, G.A. Keoleian, C. McGlade, D.N. Meehan, S. Yeh, F. You, M. Wang, A.R. Brandt, Global carbon intensity of crude oil production, Science 361 (6405) (2018) 851–853. doi.org/10.1126/science.aar6859.

[91] S. Unnasch, R. Wiesenberg, S. Tarka Sanchez, A. Brandt, S. Mueller, R. Plevin, Assessment of Life Cycle GHG Emissions Associated with Petroleum Fuels, 2009. Life Cycle Associates Report LCA-6004-3P, prepared for New Fuels Alliance, 81 pages. www.newfuelsalliance.org/NFA_PImpacts_v35.pdf.

[92] J. Sheehan, V. Camobreco, J. Duffield, M. Graboski, H. Shapouri, Life Cycle Inventory of Biodiesel and Petroleum Diesel for Use in an Urban Bus, U. S. National Renewable Energy Laboratory, Report NREL/SR-580-24089 UC Category 1503, May 1998, 1998, 286 pages. www.nrel.gov/docs/legosti/fy98/24089.pdf.

[93] G. Conan, The long-term effects of the Amoco Cadiz oil spill, Phil. Trans. R. Soc. Lon. B Biol. Sci. 297 (1087) (1982) 323–331. doi.org/10.1098/rstb.1982.0045.

[94] R. Pashaei, M. Gholizadeh, K. Jodeiri Iran, A. Hanifi, The Effects of oil spills on ecosystem at the Persian Gulf, Int. J. Rev. Life. Sci. 5 (3) (2015) 82–89. docplayer.net/65495788-The-effects-of-oil-spills-on-ecosystem-at-the-persian-gulf.html.

[95] S. Graham, Environmental Effects of Exxon Valdez Spill Still Being Felt, Scientific American, 2003, 19 December 2003, www.scientificamerican.com/article/environmental-effects-of/.

[96] Center for Biological Diversity — the Gulf Oil Spill and the Unfolding Wildlife Disaster — A Deadly Toll, 2011, 11 pages. www.biologicaldiversity.org/programs/public_lands/energy/dirty_energy_development/oil_and_gas/gulf_oil_spill/pdfs/GulfWildlifeReport_2011.pdf.

[97] I. Bociu, B. Shin, W.B. Wells, J.E. Kostka, K.T. Konstantinidis, M. Huettel, Decomposition of sediment-oil-agglomerates in a Gulf of Mexico sandy beach, Nature Sci. Rep. 9 (2019) 10071. doi.org/10.1038/s41598-019-46301-w.

[98] P. Sun, A. Elgowainy, M. Wang, J. Han, R.J. Henderson, Estimation of U.S. refinery water consumption and allocation to refinery, Fuel 221 (2018) 542–557. doi.org/10.1016/j.fuel.2017.07.089.

[99] L. Andrés, A. Zentner, J. Zentner, Measuring the Effect of Internet Adoption on Paper Consumption. World Bank, South Asia Region, Sustainable Development Dept., Policy Research Working Paper 6965, 2014. July 2014, 34 pages. documents1.worldbank.org/curated/en/375741468337759034/pdf/WPS6965.pdf.

[100] U.N. Food and Agriculture Organization (FAO), FAO Forestry Paper 129: Environmental Impact Assessment and Environmental Auditing in the Pulp and Paper Industry − 1. Background. www.fao.org/3/V9933E/V9933E01.htm.

[101] University of Minnesota, Minnesota Technical Assistance Program − Industries − Water Use in Pulp & Paper Mills, Undated. www.mntap.umn.edu/industries/facility/paper/water/.

[102] J. Roberts (Ed.), The State of the Paper Industry − Monitoring the Indicators of Environmental Performance, Report by The Environmental Paper Network, 2007, 77 pages.

[103] The Biofore Company, From forest to Paper, the story of Our Water Footprint. A Case Study for the UPM Nordland Papier Mill, August 2011, 22 pages. waterfootprint.org/media/downloads/UPM-2011.pdf.

[104] U.S. Dept. of Energy, Energy Efficiency and Renewable Energy − Industrial Technologies Program − Energy and Environmental Profile of the U.S. Pulp and Paper Industry, December 2005, 89 pages. www.energy.gov/sites/prod/files/2013/11/f4/pulppaper_profile.pdf.

[105] U.K. Environment Agency, C. Edwards, J.M. Fry, Life Cycle Assessment of Supermarket Carrier Bags: A Review of the Bags Available in 2006, Report SC030148, February 2011, 120 pages. assets.publishing.service.gov.uk/government/uploads/system/uploads/attachment_data/file/291023/scho0711buan-e-e.pdf.

[106] Organization for Economic Co-operation and Development (OECD), Global Forum on International Investment VII − the Contribution of Services to Development and the Role of Trade Liberalization and Regulation − ODI Briefing Note 1 Overview, January 2008, 17 pages. www.oecd.org/investment/globalforum/40302909.pdf.

[107] J. Pitkänen, The Environmental Impact of Service Oriented Companies, M.Sc. Thesis, Dept. of Built Environment, School of Engineering, Aalto University, Finland, 2017, 23 November 2017, 112 pages. pdfs.semanticscholar.org/6003/a7deee0ef64a6ebcf8513f0bcbc3d75fbe4a.pdf.

[108] Google, Environmental Report 2019. 66 pages. services.google.com/fh/files/misc/google_2019-environmental-report.pdf.

[109] H. Finn, Responsible Travel − Is Tourist a Dirty Word? Dated 2 May 2018 by Lea Androic. www.huckfinncroatia.com/responsible-travel/tourist-dirty-word/.

[110] U. N. World Tourism Organization (UNWTO), Tourism in the green economy, Backgr. Rep. (2012), 167 pages. www.e-unwto.org/doi/epdf/10.18111/9789284414529.

[111] A. Hares, J. Dickinson, K. Wilkes, Climate change and the air travel decisions of UK tourists, J. Transp. Geogr. 18 (3) (2010) 466−473. doi.org/10.1016/j.jtrangeo.2009.06.018.

[112] S. Gössling, P. Peeters, Assessing tourism's global environmental impact 1900−2050, J. Sustain. Tour. 23 (5) (2015) 1−21. doi.org/10.1080/09669582.2015.1008500.

[113] J. McDowall, Managing Waste in Tourist Cities, resource.co, posted 22 August 2016. resource.co/article/managing-waste-tourist-cities-11319.

[114] U. S. Dept. of Transportation, Bureau of Transportation Statistics - Summary of Cruise Ship Waste Streams, May 20, 2017 updated. www.bts.gov/publications/maritime_trade_and_transportation/2002/html/environmental_issues_table_01.html.

[115] The Global Development Research Center (GDRC), Environmental Impacts of Tourism − Tourism's Three Main Impact Areas, Undated. www.gdrc.org/uem/eco-tour/envi/one.html.

[116] U. N. Environment Programme (UNEP), A Manual for Water and Waste Management: What the Tourism Industry Can Do to Improve its Performance, 2003, 68 pages. www.unep.fr/shared/publications/pdf/WEBx0015xPA-WaterWaste.pdf.

[117]   N.W. Kuo, P.-H. Chen, Quantifying energy use, carbon dioxide emission, and other environmental loads from island tourism based on a life cycle assessment approach, J. Clean. Prod. 17 (2009) 1324–1330. doi.org/10.1016/j.clepro.2009.04.012.

[118]   U.S. Environmental Protection Agency, Energy Star Program – Data Trends – Energy Use in Retail Stores, January 2015, 2 pages. www.energystar.gov/sites/default/files/tools/DataTrends_Retail_20150129.pdf.

[119]   C.D. Whitehead, M. Melody, Water Data Report – Annotated Bibliography, Report Prepared by the Lawrence Berkeley National Laboratory for the U.S. Environmental Protection Agency WaterSense Program, 2007. May 2007, 56 pages. www.osti.gov/servlets/purl/926300.

[120]   FluidLytix – Intelligent Water Management – Homepage, visited September 2020. www.fluidlytix.com/businesses-save-water.

[121]   S. Rana, S. Pichandi, S. Karunamoorthy, A. Bhattacharyya, S. Parveen, R. Fangueiro, Carbon footprint of textile and clothing products, Chapter 7 in S.S. Muthu (Ed.), Handbook of Sustainable Apparel Production, Taylor & Francis, 2015, pp. 141–165. doi.org/10.1201/b18428-10.

[122]   A.M. Giacomin, J.B. Garcia Jr., W.F. Zonatti, M.C. Silva-Santos, M.C. Laktim, J. Baruque-Ramos, Silk industry and carbon footprint mitigation, in: Proc. 17th World Textile Conf. AUTEX. IOP Conf. Series. Materials Sci. Eng, 254, 2017, p. 192008. doi.org/10.1088/1757-899X/254/19/192008.

[123]   M.F. Ashby, Materials and the Environment – Eco-Informed Material Choice, second ed., Butterworth-Heinemann, 2013, 616 pages.

[124]   G. Peters, M. Svanström, S. Roos, G. Sandin, B. Zamani, Carbon footprints in the textile industry, in: S.S. Muthu (Ed.), Chapter 1 in Handbook of Life Cycle Assessment (LCA) of Textiles and Clothing, Elsevier, 2015, pp. 3–30. doi.org/10.1016/B978-0-08-100169-1.00001-0.

[125]   M.A. Shaikh, Water Conservation in Textile Industry, PTJ, November 2009, 4 pages. sswm.info/sites/default/files/reference_attachments/SHAKIH%202009%20Water%20conservation%20in%20the%20textile%20industry.pdf.

[126]   O EcoTextiles, Textiles and Water, Undated (Circa 2010). oecotextiles.blog/2010/02/24/textiles-and-water-use/.

[127]   Pacific Institute - Waste Not, Want Not: The Potential for Urban Water Conservation in California – Appendix F: Details of Industrial Water Use and Potential Savings, by Sector, November 2013, 31 pages. pacinst.org/wp-content/uploads/2013/02/appendix_f3.pdf.

[128]   R. Kant, Textile dyeing industry an environmental hazard, Nat. Sci. 4 (1) (2012) 22–26. doi.org/10.4236/ns.2012.41004.

[129]   Business for Social Responsibility, Apparel Industry Life Cycle Carbon Mapping, June 2009, 23 pages. www.bsr.org/reports/BSR_Apparel_Supply_Chain_Carbon_Report.pdf.

[130]   A.K. Chapagain, A.Y. Hoekstra, H.H.G. Savenije, R. Gautam, The Water Footprint of Cotton Consumption, Research Report Series No. 18, UNESCO-IHE, 2005. doi.org/10.1016/j.ecolecon.2005.11.027. September 2005, 39 pages. See also condensed version published in 2006.

[131]   A. Sule, Life cycle assessment of clothing process, Res. J. Chem. Sci. 2 (2) (2012) 87–89. citeseerx.ist.psu.edu/viewdoc/download?doi=10.1.1.1050.2964&rep=rep1&type=pdf.

Chapter 13

Making the numbers speak

As a rule, large numbers are not easily grasped. They just seem vaguely large until they can be related to something we can comprehend. Furthermore, writing for newspapers, magazines, and blogs does not permit the use of exponents because the general public is not familiar with the powers of 10, and most readers will only have a fuzzy idea of what the prefixes "mega" and "giga" stand for. There is therefore a need to express large numbers as small multiples of large yardsticks. The larger the number, the bigger the yardstick. For example, if 1 million gallons of fuel have been saved, the corresponding distance by the average US automobile (consuming 25.1 miles per gallon [1]) is 25.1 million miles, which is better expressed as 54 round trips to the moon. This chapter suggests a variety of yardsticks for several types of quantities.

13.1 Yardsticks for distances

Across the United States.

Driving distance along the US West Coast from the Canadian border to the Mexican border
    1,380 miles =2,221 kilometers
Driving distance cross country from New York to Los Angeles
    2,790 miles =4,490 kilometers
Driving distance from the northern tip of Maine to the southern tip of Florida
    3,188 miles =5,131 kilometers

Elsewhere on the planet.

Length of the Trans-Canadian Highway
    4,860 miles =5,821 kilometers
Length of the Trans-Siberian Railway (from Moscow to Vladivostok)
    5,772 miles =9,289 kilometers
Length of Australia's Highway 1 (around the entire island, the longest highway in the world)
    9,010 miles =14,500 kilometers
Earth's diameter (at the equator)
    7,926 miles =12,756 kilometers

Data, Statistics, and Useful Numbers for Environmental Sustainability.
https://doi.org/10.1016/B978-0-12-822958-3.00015-7

Earth's equatorial circumference
    24,902 miles =40,075 kilometers

Extra-planetary distances.

Earth to moon (surface to surface)
    233,811 miles =376,202 kilometers
    or approximately 30 earth diameters
Earth to sun (average over the course of the year)
    92.96 million miles =149.6 million kilometers

13.2 Yardsticks for volumes

In discussions of solid waste and water effluents, one often needs to express volumes in a meaningful way. A popular choice of unit for sizes too large to be expressed in liters or gallons is the Olympic-size swimming pool (50 m long, 25 m wide, and 2 m deep), which holds 2,500 m^3 =88,300 ft^3 =660,400 US gallons [2]. Note that some Olympic swimming pools are 3 m deep, making this yardstick somewhat ambiguous.

Alternatively, a volume can be expressed as a layer thickness covering an American football field, the dimensions of which are 300 ft × 160 ft =48,000 ft^2 =4,459 m^2 [3]. The international soccer field has no strict dimensions.

Should the volume of the Olympic swimming pool or the area of an American football field happen to be too small, a better choice is the volume of the Empire State Building in New York, which can be calculated from the total floor area (2,768,591 ft^2) multiplied by the average height of one floor (1,250 ft/102 floors =12.25 ft) and found to be 3.393 × 10^7 ft^3 =960,760 m^3 [4].

Still larger units may be taken as a certain thickness covering the continental United States with area equal to 3,201,265 square miles =8,291,238 km^2 (3,794,100 square miles =9,826,675 km^2 if including Alaska and Hawaii). The area of continental Africa is 11,668,598 square miles =30,221,532 km^2.

Yet larger volumes may be taken as those of the largest lakes on the planet, although these may mean less to the average reader as few people know whether they are deep or shallow.

Lake volumes as possible yardsticks		
Lake (location)	Volume	Surface area
Lake Baikal (Asia)	22,995 km^3	30,500 km^2
Lake Tanganyika (Africa)	17,800 km^3	32,900 km^2
Lake Superior (North America)	12,100 km^3	82,100 km^2
Lake Malawi (Africa)	7,775 km^3	30,044 km^2

Lake Michigan (North America)	4,920 km^3	57,800 km^2
Lake Huron (North America)	3,540 km^3	59,600 km^2
Lake Victoria (Africa)	2,700 km^3	69,485 km^2

Sources: [5,6].

Incidentally, the volumetric flow rate through Niagara Falls is 2,800 m^3/sec (=6 million ft^3/min) during peak daytime tourist hours [7].

13.3 Energy and carbon emissions as cars on/off the road

We often hear that certain energy conservation measures are equivalent to keeping so many cars off the road.

In the United States, the average fuel efficiency of passenger cars was 25.1 miles per gallon (9.36 L per 100 km) in 2018 [1], and the average distance driven annually was 11,576 miles (18,626 km) ([8] Table 1.8). Thus, the annual fuel consumption of a typical car is 461 gallons (1,746 L) per year. At 117 MJ per gallon of gasoline, this is equivalent to 53,960 MJ per car per year (=14,990 kWh per car per year). In terms of carbon dioxide (353 g/mi [1]), this amounts to 4,086 kg of CO_{2eq} emitted per car per year.

13.4 Energy and carbon emissions as homes

In 2018, the average electricity consumption in a US residence was 10,972 kWh/year =914 kWh/month, with Tennessee having the highest annual consumption at 15,394 kWh and Hawaii the lowest at 6,213 kWh [9]. Translated into carbon emission, the numbers are 8,350 kg of CO_{2eq} per year for the average home (11,710 and 4,730 kg CO_{2eq}/year for Tennessee and Hawaii, respectively).

13.5 Carbon footprint equivalencies

One million metric tons of carbon dioxide-equivalent emission equals [10]:

- The combustion of 530,000 short tons (=480,800 metric tons) of coal;
- The coal input of one 200 MW coal plant in about 1 year;
- The combustion of 18 billion cubic feet (=510 billion m^3) of natural gas;
- The combustion of 119 million gallons (=450 million liters) of gasoline, which is the combustion of gasoline for 7 hours in the United States, equivalent to 700,000 passenger cars each making a round trip from New York to Los Angeles;
- The combustion of 192 million gallons (=727 million liters) of LPG;
- The combustion of 107 million gallons (=405 million liters) of kerosene;

- The combustion of 102 million gallons (=386 million liters) of distillate fuel;
- The combustion of 87 million gallons (=329 liters) of residual fuel (No. 6 fuel);
- 17 minutes of world energy emissions (in 2010);
- 90 minutes of US energy emissions (in 2010);
- 3.9 hours of US buildings energy emissions;
- 7 hours of US residential buildings energy emissions;
- 8 hours of US commercial buildings energy emissions;
- 1.2 days of lighting in US buildings;
- Average annual per capita emissions of 53,000 people in the United States.

The energy released by the Tōhoku earthquake and tsunami in Japan on 11 March 2011 was estimated to be 0.19 EJ $=1.9 \times 10^{17}$ J based on its rating of 9.0 on the moment magnitude scale [11].

13.6 Paper as trees

How many times have we heard that recycling so much paper has prevented the cutting of so many trees? While there is not a single paper-to-tree conversion factor because different tree species and different manufacturing methods produce different amounts of paper, we can use the following numbers.

On average, the production of 1 metric ton of virgin paper requires 17 trees, consumes 51,500 MJ of energy, 25 m^3 of water, 680 gallons (2.57 m^3) of oil and generates 1,150 kg of CO_2. See Section 1.3 for further numbers.

The production of 1 metric ton of recycled paper saves 44% in energy, 38% in greenhouse gas emissions, 50% in water, and, of course, 100% in trees. See Section 11.5.2 for further numbers.

In other words, the recycling of 1 metric ton of paper saves 17 trees, 22,660 MJ of energy, 12.5 m^3 of water, and 437 kg of CO_{2eq}.

Sources

[1] U.S. Environmental Protection Agency (EPA), Automotive Trends Report − Highlights of the Automotive Trends Report − Figure ES-1, updated 6 March 2019. www.epa.gov/automotive-trends/highlights-automotive-trends-report.

[2] Wikipedia, Olympic-size swimming pool. en.wikipedia.org/wiki/Olympic-size_swimming_pool.

[3] Wikipedia, American Football Field. en.wikipedia.org/wiki/American_football_field.

[4] Wikipedia, Empire State Building. en.wikipedia.org/wiki/Empire_State_Building.

[5] ThoughtCo, The Largest Lakes in the World. www.thoughtco.com/largest-lakes-in-the-world-4158614.

[6] U.S. Environmental Protection Agency, The Great Lakes − Physical Features of the Great Lakes. www.epa.gov/greatlakes/physical-features-great-lakes.

[7]　Niagara Parks, Niagara Falls Geology: Facts & Features. www.niagaraparks.com/visit-niagara-parks/plan-your-visit/niagara-falls-geology-facts-figures/.

[8]　U.S. Energy Information Administration, Total Energy — Table 1.8 Motor Vehicle Mileage, Fuel Consumption, and Fuel Economy. www.eia.gov/totalenergy/data/monthly/pdf/sec1.pdf.

[9]　U.S. Energy Information Administration, Frequently Asked Questions — How Much Electricity Does an American home Use?. www.eia.gov/tools/faqs/faq.cfm?id=97&t=3.

[10]　U.S. Department of Energy, Energy Efficiency & Renewable Energy — 2011 Buildings Energy Data Book — 1.5: Generic Fuel Quad and Comparison — 1.5.3 Carbon Emissions Comparisons, on page 1-30.

[11]　O. Norio, T. Ye, Y. Kajitani, P. Shi, H. Talano, The 2011 eastern Japan great earthquake disaster: overview and comments, Int. J. Disaster Risk Sci. 2 (1) (2011) 34–42, https://doi.org/10.1007/s13753-011-0004-9.

Index

Note: 'Page numbers followed by "*f*" indicate figures and "*t*" indicate tables.'